CARBON FIBER

CARBON FIBER

SECOND EDITION

PRATIMA BAJPAI
Consultant-Pulp and Paper, Kanpur, India

ELSEVIER

Elsevier
Radarweg 29, PO Box 211, 1000 AE Amsterdam, Netherlands
The Boulevard, Langford Lane, Kidlington, Oxford OX5 1GB, United Kingdom
50 Hampshire Street, 5th Floor, Cambridge, MA 02139, United States

© 2021 Elsevier Inc. All rights reserved.

No part of this publication may be reproduced or transmitted in any form or by any means, electronic or mechanical, including photocopying, recording, or any information storage and retrieval system, without permission in writing from the publisher. Details on how to seek permission, further information about the Publisher's permissions policies and our arrangements with organizations such as the Copyright Clearance Center and the Copyright Licensing Agency, can be found at our website: www.elsevier.com/permissions.

This book and the individual contributions contained in it are protected under copyright by the Publisher (other than as may be noted herein).

Notices

Knowledge and best practice in this field are constantly changing. As new research and experience broaden our understanding, changes in research methods, professional practices, or medical treatment may become necessary.

Practitioners and researchers must always rely on their own experience and knowledge in evaluating and using any information, methods, compounds, or experiments described herein. In using such information or methods they should be mindful of their own safety and the safety of others, including parties for whom they have a professional responsibility.

To the fullest extent of the law, neither the Publisher nor the authors, contributors, or editors, assume any liability for any injury and/or damage to persons or property as a matter of products liability, negligence or otherwise, or from any use or operation of any methods, products, instructions, or ideas contained in the material herein.

Library of Congress Cataloging-in-Publication Data
A catalog record for this book is available from the Library of Congress

British Library Cataloguing-in-Publication Data
A catalogue record for this book is available from the British Library

ISBN: 978-0-12-821890-7

For information on all Elsevier publications
visit our website at https://www.elsevier.com/books-and-journals

Publisher: Susan Dennis
Acquisitions Editor: Kostas Marinakis
Editorial Project Manager: Mariana L. Kuhl
Production Project Manager: Joy Christel Neumarin Honest Thangiah
Cover Designer: Greg Harris

Typeset by SPi Global, India

Dedication

"This book is dedicated to my beloved parents and family" for their love, endless support, encouragement, and sacrifices.

Contents

List of figures   *xi*
List of tables   *xv*
Preface   *xvii*
Acknowledgments   *xix*

1. Introduction   1
   References   10
   Further reading   12

2. Raw materials and processes for the production of carbon fiber   13
   2.1 Polyacrylonitrile based carbon fibers   13
   2.2 Pitch-based carbon fibers   19
   2.3 Cellulose-based carbon fibers (CBCF)   22
   2.4 Carbon fiber from lignin   25
   2.5 Gas-phase grown carbon fibers   36
   2.6 Carbon nanotubes (CNTs)   37
   2.7 Other carbon fiber precursors   38
   2.8 Safety concerns   40
   References   41
   Further reading   50

3. Types of lignins and characteristics   51
   3.1 Lignosulfonates   55
   3.2 Kraft lignin   56
   3.3 Soda lignin   58
   3.4 Steam explosion lignin   58
   3.5 Organosolv lignins   59
   References   63
   Further reading   66

4. Conversion of lignin to carbon fiber   67
   4.1 Melt spinning   67
   4.2 Dry spinning   67
   4.3 Wet spinning   68
   4.4 Dry-jet wet spinning   68

    4.5 Electrospinning — 69
    References — 74
    Further reading — 75

5. Properties of carbon fiber — 77
    References — 88
    Further reading — 89

6. Recycling of carbon fiber-reinforced polymers — 91
    6.1 Methods for recycling — 94
    6.2 Mechanical recycling — 95
    6.3 Energy consumption of the recycling processes — 116
    6.4 Energy and environmental benefits at the right price — 118
    References — 118
    Relevant websites — 123
    Further reading — 123

7. Manufacturing of composites from recycled carbon fiber — 125
    References — 136
    Further reading — 138

8. Applications of carbon fiber/carbon fiber-reinforced plastic/recycled carbon fiber-reinforced polymers — 139
    8.1 Applications — 140
    References — 152
    Relevant websites — 155
    Further reading — 155

9. The carbon fiber/carbon fiber-reinforced plastic/recycled carbon fiber-reinforced polymer market — 157
    9.1 Markets for recycled carbon fiber — 163
    References — 168
    Relevant websites — 170
    Further reading — 170

10. Future directions of the carbon fiber industry — 171
    References — 180
    Relevant websites — 181

11. Future research on carbon fibers — 183
References — 186
Relevant websites — 186
Further reading — 186

Appendix: Carbon nanotubes — *187*
Glossary — *213*
Index — *217*

List of figures

Fig. 1.1	Carbon fiber demand over the years.	5
Fig. 2.1	Chemical structure of (A) polyacrylonitrile and (B) microstructure of MP carbon fibers.	14
Fig. 2.2	Process for manufacture of carbon fiber from polyacrylonitrile.	15
Fig. 2.3	Possible reactions occurring during the carbonization of stabilized polymer polyacrylonitrile precursor.	17
Fig. 2.4	Scanning electron micrographs of polyacrylonitrile-based carbon fibers: (A) low and (B) high magnification.	18
Fig. 2.5	Process for manufacture of carbon fiber from pitch.	19
Fig. 2.6	Possible reaction mechanisms of the oxidative stabilization of pitch precursor. (a) NP80; (b) NHP; and (c) A60	21
Fig. 2.7	Scanning electron micrographs of pitch-based carbon fibers: (A) low and (B) high magnification.	22
Fig. 2.8	Scanning electron micrographs of rayon-based carbon fibers: (A) low and (B) high magnification.	24
Fig. 2.9	Scanning electron microscopy (SEM) micrographs of cellulose-based carbon fibers (CBCFs): (A) and (B) side surface, (C) and (D) crossing section.	25
Fig. 2.10	Scanning electron microscopy images of (A and B) lignin fibers extruded from an organic solvent-extracted kraft hardwood lignin; (C and D) oxidized lignin fiber at heating rates of (C) $0.05°C\ min^{-1}$ and (D) $0.025°C\ min^{-1}$ showing different degrees of fusion; (E and F) carbon fiber, carbonized to 1000C at $2°C\ min^{-1}$ after stabilization at a heating rate of $0.01°C\ min^{-1}$.	32
Fig. 3.1	Monolignol monomer species. (A) p-Coumaryl alcohol (4-hydroxyphenyl [H]), (B) coniferyl alcohol (guaiacyl [G]), (C) sinapyl alcohol (syringyl [S]).	52
Fig. 3.2	Chemical structure of hardwood and softwood lignin.	53
Fig. 3.3	Photographs showing the differences in physical appearance for a number of typical technical lignins.	62
Fig. 4.1	Production steps involved in the production of carbon fiber from lignin.	70
Fig. 6.1	Main technologies for CFRP recycling. (A) Mechanical recycling. (B) Fiber reclamation.	95

Fig. 6.2	Scanning electron microscopy of recycled (through pyrolysis) carbon fibers. (A) Clean recycled fibers. (B) Recycled fibers with char residue.	99
Fig. 6.3	Adherent Technologies Inc.'s recycled CF derived from the catalytic conversion process.	100
Fig. 6.4	ATI's Phoenix reactor, a pilot scale vacuum pyrolysis unit.	102
Fig. 6.5	Low-temperature, low-pressure recycling reactor.	103
Fig. 6.6	Schematic of carbon fiber-reinforced polymer (CFRP) waste treatment by microwave thermolysis.	104
Fig. 6.7	The macroscopic appearances of the CFRP waste samples after experiments (A) 30 min of traditional heating at 400°C, (B) 30 min of traditional heating at 450°C, (C) 30 min of traditional heating at 500°C, and (D) 13 min of microwave heating at 450°C.	104
Fig. 6.8	Product range of *ELG Carbon Fibre Ltd*.	107
Fig. 6.9	Recycling furnace (ELG carbon Fibre Ltd.).	107
Fig. 6.10	Fluidized bed method.	108
Fig. 6.11	Supercritical fluid method.	111
Fig. 9.1	A map of carbon fiber manufacturing facilities.	158
Fig. 9.2	Examples of demonstrators manufactured with recycled CFs: (A) Wing mirror covers (BMC compression, Warrior et al., 2009), (B) Aircraft seat arm-rest (3-DEP process, George, 2009), and (C) Rear or WorldFirst F3 car (woven re-impregnation, Meredith, 2009).	165
Fig. A.1	Conceptual diagrams of single-walled carbon nanotubes (SWCNT) (A) and multiwalled carbon nanotubes (MWCNT) (B).	188
Fig. A.2	A carbon nanotube with closed ends.	188
Fig. A.3	Carbon nanotube structures of armchair, zigzag, and chiral configurations. They differ in chiral angle and diameter: armchair carbon nanotubes share electrical properties similar to metals. The zigzag and chiral carbon nanotubes possess electrical properties similar to semiconductors. (A) Armchair. (B) Zigzag. (C) Chiral.	189
Fig. A.4	Idealized models of (A) zigzag and (B) armchair monolayer nanotubes.	190
Fig. A.5	Carbon nanotubes arc discharge production method.	193
Fig. A.6	Schematic synthesis apparatus. (A) Classical laser ablation technique. (B) Ultrafast laser evaporation (free electron laser (FEL)).	194
Fig. A.7	Catalyzed chemical vapor deposition.	195
Fig. A.8	The SEM images of CNTs (A) after (B) before purification stages with HCl.	204

Fig. A.9 Scanning electron microscopy (SEM) micrographs of the pristine nanotubes (PNTs) (A) and cut nanotubes (CNTs) (B). After the mechanical treatment the general aspect of the surface is clearly modified, and nanotubes with short lengths are observed. 205

Fig. A.10 Typical transmission electron microscopy micrograph of multiwalled carbon nanotubes. 206

List of tables

Table 1.1	Properties of carbon fiber.	2
Table 1.2	Advantages of carbon fiber-reinforced carbon composites.	2
Table 1.3	Application area of carbon fiber.	3
Table 1.4	Global carbon fiber demand by application in 1000 tons (2013) (total 46,500 t).	3
Table 1.5	Estimated global carbon fiber consumption.	4
Table 1.6	Market share of carbon fibers depending on precursor type.	6
Table 1.7	Repartition of the production costs of PAN-based carbon fibers.	9
Table 1.8	Advantages of lignin over other precursors like MPP and PAN.	10
Table 2.1	Advantages of extruding lignin rather than polyacrylonitrile.	26
Table 2.2	Properties of carbon fiber from different types of lignin.	28
Table 2.3	Physical properties of carbon fibers from blends of lignin-PEO.	30
Table 2.4	Physical properties of carbon fibers from blends of lignin.	30
Table 2.5	Mechanical properties of carbon fibers made from lignin.	31
Table 3.1	The characterization of technical lignins.	54
Table 3.2	Characteristics of the technical lignin.	55
Table 3.3	Some of the major manufacturers of lignins.	55
Table 3.4	Sulfur content and purity of different types of lignins.	62
Table 4.1	Most common spinning techniques.	68
Table 5.1	Comparison of carbon fibers properties with other materials.	78
Table 5.2	Axial tensile properties of carbon fibers.	79
Table 5.3	Prices of carbon fiber of different categories.	80
Table 5.4	Classification of carbon fiber.	81
Table 5.5	Mechanical properties of different types of carbon fibers.	82
Table 5.6	Mechanical properties of cellulosic fibers.	82
Table 5.7	Properties of carbon fibers from different precursors.	83
Table 5.8	Typical structural parameters for the selected pitch- and PAN-based carbon fibers.	84
Table 5.9	Properties of commercial carbon fibers.	85
Table 5.10	Advantages of carbon fiber-reinforced carbon composites.	87
Table 5.11	Properties of various engineering fibers.	87
Table 5.12	Thermal conductivity and electrical conductivity of carbon fiber and carbon nanotubes.	87

Table 6.1	Advantages of recycled carbon fibers.	92
Table 6.2	Carbon fiber recycling industry progress.	94
Table 6.3	Grades of sheet molding compound recyclate from ERCOM GmbH.	96
Table 6.4	Grades of sheet molding compound recyclate from Phoenix Fibreglass, Inc.	97
Table 6.5	Tensile properties of carbon fiber.	102
Table 6.6	Product range of *ELG Carbon Fiber Ltd.*	106
Table 6.7	Fluidized bed method.	110
Table 6.8	Supercritical fluid method.	111
Table 6.9	Comparison between existing recycling methods and the new EHD method.	114
Table 6.10	Estimated energy consumption of the main recycling processes.	116
Table 7.1	Composite materials manufactured with rCF for mechanical testing.	130
Table 7.2	Real part demonstrator and commercially available semiproducts made of rCF.	131
Table 7.3	Benefits and problems with different processes used for remanufacturing of composites from recycled carbon fibers.	136
Table 8.1	Current applications of carbon fiber composites.	140
Table 8.2	Global carbon fiber share by application in 2017.	146
Table 9.1	Carbon fiber manufacturing cost breakdown Rocky Mountain Institute (2015).	158
Table 9.2	Manufacturers of carbon fiber.	159
Table 9.3	Worldwide production capacities of carbon fibers (metric tons annually).	159
Table 9.4	Production units of different producers in different regions of the world.	160
Table 9.5	Demonstrators manufactured with recycled CFs.	165
Table 9.6	Potential structural applications for rCFRPs.	166
Table 10.1	Future directions of carbon fiber industry.	172
Table A.1	Properties of CNTs.	191
Table A.2	Tabular representation of CNT properties.	191

Preface

Carbon fibers are state-of-the-art materials with superior mechanical properties including high-specific strength and high-specific modulus and characteristics such as low density, low thermal expansion, heat resistance, and chemical stability. In addition, various kinds of carbon fibers with differing fiber morphology or mechanical performance are developed. With the characteristics mentioned earlier, carbon fibers are applied to various fields. Carbon fibers have been extensively used in composites in the form of woven textiles, prepregs, continuous fibers/rovings, and chopped fibers. The composite parts can be produced through filament winding, tape winding, pultrusion, compression molding, vacuum bagging, liquid molding, and injection molding. In recent years the carbon fiber industry has been growing steadily to meet the demand from different industries. Most of the carbon fiber manufacturers have plans for expansion to meet the market demand. However, the large-volume application of carbon fiber in automotive industry has been hindered due to the high fiber cost and the lack of high-speed composite fabrication techniques. The current carbon fiber market is dominated by polyacrylonitrile carbon fibers, while the rest is pitch carbon fibers and a very small amount of rayon carbon fiber textiles. Different precursors produce carbon fibers with different properties. Global demand for carbon fiber is forecast to grow to 140,000 tonnes by 2020. Carbon fiber-reinforced polymer (CFRP) is a light fiber-reinforced polymer that is incredibly strong, composed of carbon fibers. These composites are highly desired and high value materials exhibiting superior strength to weight properties. Polyepoxide (epoxy) is the polymer used most of the time, but other polymers like vinyl, nylon, ester, or even polyester can be used in some cases. Other fibers like glass fibers, Kevlar, and aluminum may also be used along with carbon fibers.

Today a small fraction of the carbon fiber composite materials used is recycled. However, new legislation polices and approaching shortage of raw materials, in combination with the ever increasing use of carbon fiber composites, force society to recycle these materials in the near future. The objective of this book is to bring together available information on the production, properties, application, and future of carbon fibers. This will be of interest to those involved in the investigation of carbon fiber, carbon fiber manufacturers, and the users of carbon fibers. This book will also be of interest to those involved in the recycling of carbon fiber-reinforced polymers and manufacturing of composites from recycled carbon fiber-reinforced polymers. Students engaged in the field of chemistry, materials science, and polymer science will also find this book very useful.

Acknowledgments

I am grateful for the help of many people, companies, and publishers for providing information and granting permission to use their material. Deepest appreciation is extended to Elsevier, Springer, Wiley, Royal Society of Chemistry, Hindawi, MDPI, and other open-access Journals and publications for all the information used in this report.

Some excerpts taken from Bajpai P. Carbon fibre from lignin, Springer Briefs in material science. Springer; 2017.

Some excerpts taken from Holmes M. Global carbon fibre market remains on upward trend. Reinf Plast 2014;58:38–45.

Some excerpts taken from Xiaosong H. Fabrication and properties of carbon fibers. Materials 2009;2:2369–2403.

Some excerpts taken from Zeng F, Pan D, Pan N. Choosing the impregnants by thermogravimetric analysis for preparing Rayon-based carbon fibers. J Inorg Org Polym Mater 2005;15:261–267.

Some excerpts taken from Chanzy H, Paillet, M, Hagege, R. Spinning of cellulose from N-methylmorpholine N-oxide in the presence of additives. Polymer 1990;31:400–405.

Some excerpts taken from Souto F, Calado V, Pereira N. Lignin-based carbon fiber: a current overview. Mater Res Express 2018;5:072001.

Some excerpts taken from Baker DA, Rials TG. Recent advances in low-cost carbon fibre manufacture from lignin. J Appl Polym Sci 2013;130:713–28.

Some excerpts taken from Yokoyama A, Nakashima N, Shimizu K. A new modification method of exploded lignin for the preparation of a carbon fiber precursor. J Appl Polym Sci 2003;48:1485–1491. https://doi.org/10.1002/app.1993.070480817.

Some excerpts taken from Matsushita Y. Conversion of technical lignins to functional materials with retained polymeric properties. Wood Sci 2015;61:230–250. https://doi.org/10.1007/s10086-015-1470-2.

Some excerpts taken from Niaounakis M. Biopolymers: Applications and trends. 1st ed. 2015.

Some excerpts taken from Li J, Henriksson G, Gellerstedt G. Lignin depolymerization and its critical role for delignification of aspen wood by steam explosion. Biores Technol 2007;98:3061–8.

Some excerpts taken from Minus ML, Kumar S. Carbon fibre. Kirk-Othmer Encycl Chem Technol 2007;26:729–749.

Some excerpts taken from Zhu JH, Chen Pi-yu, Su M, Peia C, Xing F. Recycling of carbon fibre reinforced plastics by electrically driven heterogeneous catalytic degradation of epoxy resin. Green Chem 2019;21:1635–1647.

Some excerpts taken from Pickering SJ. Recycling technologies for thermoset composite materials—current status. Compos A Appl Sci Manuf 2006;37:1206–1215.

Some excerpts taken from Pimenta S, Pinho ST. Recycling carbon fibre reinforced polymers for structural applications: technology review and market outlook. Waste Management 2011;31(2):378–392.

Some excerpts taken from Nahil MA, Williams PT. Recycling of carbon fibre reinforced polymeric waste. J Anal Appl Pyrol 2011;91(1):67–75.

Some excerpts taken from Pickering SJ, Liu Z, Turner TA, Wong KH. Applications for carbon fibre recovered from composites. IOP Conf Ser Mater Sci Eng 2016;139: 1–18.

Some excerpts taken from Lester E, Kingman S, Wong, KH, Rudd C, Pickering S, Hilal N. Microwave heating as a means for carbon fibre recovery from polymer composites: a technical feasibility study. Mater Res Bull 2004;39(10):1549–1556.

CHAPTER 1

Introduction

Carbon fiber—*"the world's structural wonder material"*—is also known as graphite fiber. It is very lightweight and very strong material. In comparison with steel, it is five times stronger and two times stiffer and lighter, which makes it an excellent manufacturing material for several parts. Engineers and designers prefer carbon fiber for manufacturing. The reasons are presented in Table 1.1.

These fibers are not used as such. They are used to reinforce materials such as epoxy resins and other thermosetting materials. These materials are termed as composites as they possess more than one component and are quite strong for their weight. They are very strong in comparison with steel but very much lighter. Due to this property, they may be used to replace metals in several applications, from parts for airplanes and the space shuttle to tennis rackets and golf clubs.

Table 1.2 shows advantages of carbon fiber-reinforced carbon composites.

Carbon fibers date back to 1879. Thomas Edison baked cotton threads or bamboo silvers at elevated temperatures, which carbonized them into an all-carbon fiber filament. In 1958 Roger Bacon of Ohio in Cleveland, Ohio, United States, produced high-performance carbon fibers. Leslie Philips, a British engineer, in 1964, realized the high strength of carbon fiber. Later on, carbon fibers produced from rayon strands processed by carbonation were developed. Akio Shindo in the early 1960s produced carbon fibers from PAN. For manufacturing PAN-based carbon fibers, PAN is processed to a fibrous shape by spinning and then subjected to oxidation, carbonization, and surface treatment. Leonard Singer produced carbon fibers from pitch in 1970. The manufacturing process involves in making petroleum or coal pitch into a fibrous shape; then oxidation, carbonization, and surface treatment are performed (Saito et al., 2011). These fibers were not efficient and contained about 20% carbon. The strength and stiffness properties were inferior. The US Air Force and NASA started using carbon fiber in aircraft and spacecraft application.

> *During the 1970s, work was conducted to find alternative raw materials for the production of carbon fibers made from a petroleum pitch obtained from oil processing. These fibers contained about 85% carbon and possessed excellent flexural strength. But, they had very little compression strength and were not very much accepted.*
>
> *(www.madehow.com)*

Table 1.1 Properties of carbon fiber.

High strength-to-weight ratio
Good rigidity
Corrosion resistant
Electrically conductive
Fatigue resistant
Good tensile strength but brittle
Fire resistance/not flammable
High thermal conductivity
Low coefficient of thermal expansion and low abrasion
Nonpoisonous
Biologically inertness and X-ray permeability
Self-lubricating
Excellent electromagnetic interference shielding property
Relatively expensive
Requires specialized experience and equipment to use
High dimensional stability, low coefficient of thermal expansion, and low abrasion
High damping
Electromagnetic properties

http://www.innovativecomposite.com/what-is-carbon-fiber/; https://www.motioncomposites.com/en/about-carbon-fiber/

Table 1.2 Advantages of carbon fiber-reinforced carbon composites.

Resistance to high temperatures and weathering, low flammability, low smoke density, low toxicity of decomposition products. Temperature resistance of course depends on choice of resin
High chemical stability
Large variety of possible component shapes and sizes
High durability due to long prepreg storage life
Prepregs comprise the range of reinforcements and resin matrix combinations. They are manufactured on a state-of-the-art fusible resin plant. Fusible resins have fewer volatile constituents and increase the composite materials' mechanical strength. The prepreg manufacturing plant is accredited to DIN ANDISO 9001 quality assurance standards

Nowadays, carbon fibers have become an important part of several products, and new applications are being developed. The leading producers of carbon fibers are the United States, Japan, and Western Europe.

Worldwide carbon fiber is in rapidly growing demand as a lightweight and strong alternative to metal for various industries such as aeronautics, automotive, marine, transportation, construction, electronics, and wind energy (Fitzer et al., 1989; Hajduk, 2005;

Huang, 2009; Saito et al., 2011; Barnes et al., 2007; Soutis, 2005; Ogawa, 2000; Nolan, 2008; van der Woude et al., 2006; Fuchs et al., 2008; Zhang and Shen, 2002; Aoki et al., 2009; Tran et al., 2009; Olenic et al., 2009; Baughman et al., 2002; Thostensona et al., 2001; Roberts, 2006; Todd, 2019; Figueiredo et al., 1990; Chung, 1994; Watt, 1985; Donnet and Bansal, 1990; Minus and Kumar, 2005, 2007). Carbon nanofibers have been explored for use in regenerative medicine and also for treatment of cancer (Ogawa, 2000; Nolan, 2008; van der Woude et al., 2006; Fuchs et al., 2008. Zhang and Shen, 2002; Aoki et al., 2009; Barnes et al., 2007; Soutis, 2005; Tran et al., 2009; Olenic et al., 2009; Baughman et al., 2002; Thostensona et al., 2001). Table 1.3 shows the application of carbon fiber (Holmes, 2014). Applications in aerospace and defense, sport/leisure sector, and wind turbines have grown substantially. The automotive segment is also becoming very important. This could be because of the ramp-up phase for the production of the i-models from BMW. Other applications are construction of molding and compounding plant, pressure vessels, civil engineering, and marine. Table 1.4 shows global carbon fiber demand by application in 1000 tons (2013).

Table 1.3 Application area of carbon fiber.

Polyacrylonitrile aerospace/high-end carbon fiber:
  Toray (largest worldwide manufacturer)
  TohoTenax
  Mitsubishi
  Hexcel
  Cytec
  Nippon Graphite Fiber Corporation
Polyacrylonitrile commercial grade carbon fiber:
  Zoltek
  SGL

http://www.formula1-dictionary.net/carbon_fiber.html

Table 1.4 Global carbon fiber demand by application in 1000 tons (2013) (total 46,500 t).

Aerospace and defense	13.9, 30%
Molding and compound	5.5, 12%
Sports and leisure	6.4, 14%
Wind turbines	6.7, 14%
Automotive	5.0, 11.0%
Pressure vessels	2.4, 5%
Civil engineering	2.3, 5%
Marine	0.8, 2%
Others	0.5, 7%

Based on Holmes M. Global carbon fibre market remains on upward trend. Reinf Plast 2014;58:38–45.

The diameter of carbon fiber is about 0.0002–0.0004 in. and contains at least 90% carbon by weight. It is a long, thin strand of material (Figueiredo et al., 1990; Chung, 1994; Watt, 1985; Donnet and Bansal, 1990; Minus and Kumar, 2005). The carbon atoms are bonded together in microscopic crystals. These are more or less aligned parallel to the long axis of the fiber. The crystal alignment makes the fiber very strong for its size (Chung, 1994). Thousands of carbon fibers are twisted together to form a yarn. This can be used by itself or woven into a fabric. The yarn or fabric is blended with epoxy and wound or molded into shape for producing different types of composite materials. Carbon fiber-reinforced composites are being used for making aircraft and spacecraft parts, racing car bodies, golf club shafts, bicycle frames, fishing rods, automobile springs, sailboat masts, and several other components where light weight along with high strength is required.

Several types of carbon are available. They can be sort or continuous. These are crystalline, amorphous, or partly crystalline. Table 1.5 shows estimated global carbon fiber consumption (Roberts, 2006; Red, 2006; Pimenta and Pinho, 2011).

"Demand of Carbon fiber increased from 26,500 tons in 2009 to 63,500 tons in 2016, which yielded a revenue up about US$ 2.34 billion (a growth of 8.7% related to the year 2015). According to market trends it is expected an annual growth rate between 10% to 13% for the coming years," (Fig. 1.1) (Souto et al., 2018; Witten et al., 2015, 2016, 2017).

Demand for carbon fibre would reach 89,000 t by 2020 and would generate revenues of over US$3.3 billion.

(Holmes, 2014)

Generally, mechanical properties are used to classify their material (Minus and Kumar, 2005; Chen, 2014; Hedge et al., 2004). It should be mentioned that diameter and morphology are also an important criteria for classifying the fibers when it confers important prominence on mechanical properties (Minus and Kumar, 2005).

Carbon fibers were first produced in the 1950s as a reinforcement for high-temperature molded plastic components on missiles (Figueiredo et al., 1990; Chung, 1994; Watt, 1985; Donnet and Bansal, 1990; Minus and Kumar, 2005). These were produced by heating strands of rayon till they get carbonized. The resulting fibers contained ~20% carbon and had lower strength properties. Hence, this process proved to be

Table 1.5 Estimated global carbon fiber consumption.

	2004 (tons)	2006 (tons)	2008 (tons)	2010 (tons)
Industrial	11,400	12,800	15,600	17,500
Aerospace	5600	6500	7500	9800
Sporting goods	4900	5900	6700	6900
Total	21,900	25,200	29,800	34,200

Based on Roberts T. The carbon fibre industry: global strategic market evaluation 2006–2010. Watford: Materials Technology Publications; 2006. pp. 10, 93–177, 237; Red C. Aerospace will continue to lead advanced composites market in 2006. Compos Manuf 2006;7:24–33; Pimenta S, Pinho ST. Recycling carbon fibre reinforced polymers for structural applications: technology review and market outlook. Waste Manag 2011;31:378–392.

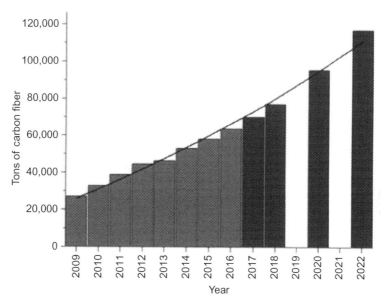

Fig. 1.1 Carbon fiber demand over the years. *(Reproduced with permission Souto F, Calado V, Pereira Jr N. Lignin-based carbon fiber: a current overview. Mater Res Express 2018;5:072001. https://doi.org/10.1088/2053-1591/aaba00.)*

inefficient. The carbon fibers were used successfully on a commercial scale in the early 1960s, as the need of the aerospace industry—especially for military aircraft—for better and lightweight materials became very important. In this process, polyacrylonitrile was used as a starting material. This process produced a carbon fiber that had very good strength properties. Therefore the polyacrylonitrile process rapidly became the main method for manufacturing carbon fibers. "Carbon fibres are being used widely in commercial and civilian aircraft, recreational, industrial, and transportation markets. Carbon fibres are used in composites with a lightweight matrix. Carbon fibre composites are particularly suited for applications where strength, stiffness, lower weight, and outstanding fatigue characteristics are the main requirements. They can be also used where high temperature, chemical inertness and high damping are important" (www.carbonfiber-vinyl.com). Carbon fibers offer 10 times the strength of steel at a quarter of the weight (Xiaosong, 2009).

Carbon fibers manufactured from polyacrylonitrile have better mechanical and physical properties in comparison with rayon-based ones. Today, they are the most promising raw materials for production of high-strength carbon fibers. Carbon fibers based on mesophase pitch turned out as more expansive following a complicated and complex process of conversion of cheap pitches into mesophase-forming modification. Table 1.6 shows market share of carbon fibers depending on the precursor type (Fitzer and Heine, 2019).

Table 1.6 Market share of carbon fibers depending on precursor type.

PAN—carbon yield 50%; middle/high modulus; high/very high strength
Market share, 95%
Mesophase pitch—carbon yield 80%; very high modulus; low/middle strength
Market share <5%
Cellulose—carbon yield 25%; poor because internal defect structure
Market share, minimal
Lignin—carbon yield 20%; poor because internal defect structure
Market share, minimal

Based on Fitzer E, Heine M. Carbon fibers in book: Industrial carbon and graphite materials—Raw materials, production and applications. [in print] Wiley-VCH; 2019.

Applications of carbon fiber composites take advantage of its strength, stiffness, low weight, fatigue characteristics, lack of corrosion, and heat insulation. Large quantities of carbon fiber are used by modern aircrafts, engineering industries, and producers of sporting goods (Carson, 2012; Hajduk, 2005). Oak Ridge National Laboratory (ORNL) in the United States has estimated a replacement of more than 60% of the ferrous metals in automobiles by carbon fiber composites. This could reduce the overall weight by about 40% and the fuel consumption by more than 20%. Now a total of about 50,000 tons of carbon fiber is being used worldwide annually, and the market is increasing by about 10% every year; Japan and the United States are the major producers. Toray Industries has plans to spend Y45 billion to increase its global carbon fiber production capacity to better cope with higher demand from growing use of the material in aircraft, autos, and other products (http://www.marketwatch.com/story/toray-to-boost-carbon-fibre-production-2012-03-08). Japan's largest synthetic fiber maker by sales said it is planning an additional 6000-ton boost in its carbon fiber output capacity per year. As a result the amount of Toray's annual production capacity is expected to reach 27,100 tons on an annual basis by March 2015 at its facilities in Japan, South Korea, the United States, and France. Currently, Toray has a combined annual carbon fiber production capacity of 17,900 tons. Polymeric materials that leave a carbon residue and do not melt upon pyrolysis in an inert atmosphere are usually considered candidates for production of carbon fiber (Figueiredo et al., 1990; Chung, 1994; Watt, 1985; Donnet and Bansal, 1990). More than 90% of the carbon fiber is produced by heat treatment and pyrolysis of polyacrylonitrile, a synthetic material produced from petroleum feedstock. Nevertheless, petroleum- or coal-based pitch or viscose fiber can also be used. According to a report by Textiles Intelligence, Wilmslow, United Kingdom, "worldwide demand for lightweight, strong carbon fibres is expected to reach 8.5 million tons in 2015, driven by interest in carbon fibre reinforced plastics for use on cars and light vehicles" (http://specialtyfabricsreview.com/articles/0910_sw14_carbon_fibres.html).

CFRP is used in aircraft, sports equipment and racing cars, where excess weight means higher fuel consumption and reduced speeds. Electric vehicles manufacturers will be the first to demand CFRP, as the lighter the car's weight, the further it goes without a recharge. A joint venture between, SGL group Charlotte, Norh Carolina, a producer of carbon fibres, and BMW, Leipzig, Germany, involved investing $100 million in a new carbon fibre plant at Moses Lake, Washington,

(specialtyfabricsreview.com)

Carbon fiber-reinforced polymer or carbon fiber-reinforced plastic is highly desired, high value materials showing superior strength-to-weight properties (Figueiredo et al., 1990; Chung, 1994; Watt, 1985; Donnet and Bansal, 1990). Polyepoxide (epoxy) is the polymer mostly used, but other polymers such as vinyl, nylon, ester, or even polyester can be used in few cases. Other fibers such as glass fibers, Kevlar, and aluminum can also be used along with carbon fibers. Carbon fiber can be classified in different ways, such as by raw materials and mechanical properties. Raw materials can be divided into polyacrylonitrile carbon fiber, pitch carbon fiber, and rayon carbon fiber. Currently the majority of structural materials are using polyacrylonitrile carbon fiber. Mechanical properties are divided as follows (Donnet and Bansal, 1990; Minus and Kumar, 2005, 2007):

(1) ultrahigh modulus (UHM) carbon fiber
(2) high modulus (HM) carbon fiber
(3) ultrahigh strength (UHS) carbon fiber
(4) high strength (HS) carbon fibers

Carbon fiber composites have remained an elusive material in the automotive industry (Red, 2006). These have been used in jet fighters and high-end race cars for more than 20 years, so there is very little doubt about its ability to build lighter and more durable vehicles. Carbon fiber offers a weight savings of 75% over steel. These fibers give sport cars a real benefit in acceleration and top speed and enable all automobiles for obtaining improved fuel economy.

"Since 1960, the carbon fibre business has grown in volume. The main markets that have driven this growth are the following industry:
- Aerospace wind energy
- Compressed natural gas storage
- Civil engineering" (www.plasticbiz360.com)

During the last decade, several important changes have taken place in carbon fiber industry. During 2005–06, there was a shortage of carbon fiber. But the market situation rapidly changed because of the economic downturn worldwide. The demand dropped in first three quarters of 2009 as the economy worsened worldwide. At the same time, supply increased that some producers had sufficient supply of carbon fibers with lower lead times to consumers. "The global carbon fibre market has grown at double digits during the last five years and would grow at a slower speed, increasing from 2008's level of USD 1.5 billion to USD 2.5 billion in 2014, with the overall market appearing positive for the next twenty years. The aerospace and defense sector represented 23% of total world carbon fibre demand by 2014 in terms of carbon fibre consumption. The aerospace industry is growing at a compound annual growth rate of 19 %, with a paradigm shift towards demand for carbon fibre for manufacturing lighter weight and high resistant aircrafts. Besides aerospace and defense, other industrial sectors are also growing though several are, by comparison, still considered niche markets. The industrial sector (excluding aerospace and sports/leisure) grew to 65% by 2014. Other largest growth area for carbon fibre is in filament wound composites for offshore oil and gas industry. The worldwide actual carbon fibre output reached 76,790 tons by 2014 and it is predicted that the offshore industry alone could require 50,000 tons by 2020. The manufacture of wind turbine blades is the fastest growing application for carbon fibre usage in the industrial market sector over the next seven years." (www.plasticbiz360.com).

Carbon fibers have also been used as biomaterials since the late 1970s. The most important benefits of using carbon fibers as biomaterials are their mechanical properties (lightweight, high strength, and flexibility). They are easy to combine with conventional biomaterials and have several fiber morphologies and high radiolucency. Carbon fibers also have required good biocompatibility and have good in vivo (in the body) stability, which are the important characteristics of biomaterials. For taking benefit of these properties, studies of clinical applications of carbon fibers for reinforcement of traditional biomaterials and to use carbon fibers alone for reinforcing the tendon were conducted. Currently, spinal fixation cages and other devices using carbon fibers are mostly applied clinically, but the previous race in development had reduced, and development of the application of carbon fibers to biomaterials had not been much active in the last decade. But due to the new developments, carbon fibers of nanosized diameter have been developed, and nanolevel control of the structures has become possible. This opens new opportunities of application to biomaterials (Tran et al., 2009; Aoki et al., 2009; Olenic et al., 2009).

The major disadvantage is the high production cost, which is limiting the supply, in spite of a growing demand. The major processing steps for the production of carbon fiber usually include spinning, stabilization, carbonization, and sometimes graphitization (Bahl et al., 1998). Nowadays the polyacrylonitrile precursor represents about half of the production costs (Table 1.7), and the equipment cost is about one-third of the production cost (Warren et al., 2009).

A large number of industries are interested in carbon fibers as a new and lightweight material with the potential to replace, for instance, the steel in cars and the glass fibers in blades in wind power stations. The precursors for producing carbon fiber are polyacrylonitrile, petroleum pitch, and regenerated cellulose (rayon) (Forrest et al., 2001). "Regarding the properties of precursors, lignin is most similar to petroleum pitch. Lignin is an aromatic biopolymer. It is amorphous, highly branched polyphenolic macromolecule with a complex structure, and the material typically forms about 1/3 of the dry mass

Table 1.7 Repartition of the production costs of PAN-based carbon fibers.

Precursor 51%
Labor 10%
Depreciation 12%
Utilities 18%
Others 9%

Based on Warren CD, Paulauskas FL, Baker FS, Eberle CC, Naskar A. Development of commodity grade, lower cost carbon fibre—commercial applications. SAMPE J 2009;45 (2):24–36.

of woody materials. It can be used as a precursor for the production of carbon fibres. It is derived from wood and plants and has a significant potential cost advantage over even textile-grade polyacrylonitrile as a precursor material for production of low-cost carbon fibre" (www1.eere.energy.gov/vehiclesandfuels/pdfs/lm/7_low-cost_carbon_fibre.p).

> Low cost replacement of lignin is expected to save 37–49% in the final production cost of carbon fibre. If polyacrylonitrile is replaced by lignin for wind turbine blades. it has a triple pay-off: it uses renewable raw materials, optimizes energy and materials costs for the production of carbon fibres, and the fibres themselves will be used in wind turbines for producing renewable energy. Lignin is a breakable biopolymer, it cannot be spun, stretched/aligned, and spooled into fibres without an modification. Lignin is a highly abundant biopolymeric material (second only to cellulose) and is presently obtained from chemical pulping of wood. As a by-product of the pulp and paper industry, approximately 50 million tons of lignin are produced annually worldwide, of which 98% to 99% of the industrial lignin is incinerated to produce steam and energy, while 1% to 2%, derived from the sulfite pulp industry, is used in chemical conversion to produce lignosulfonates.
>
> *(Bajpai, 2017; Mohan et al. 2006; Gellerstedt et al. 2012; Inwood, 2014; Norgren and Edlund, 2014)*

> In the recent years many biomass refineries are coming on stream, so, the lignin generated as a byproduct from production of cellulosic ethanol will represent a valuable material for production of carbon fibre. Lignin produced using Oganosolv pulping process can be readily melt-spinnable as isolated. These lignins are very pure in comparison to lignins produced using the chemical pulping process.
>
> *(Bajpai, 2017)*

Lignin is a renewable raw material and is also sustainable. As compared with polyacrylonitrile, the cost of manufacturing is reduced by more than 50% when lignin is used as precursor. Lignin is available in huge amounts and is inexpensive. It is a by-product of the pulp industry and biorefineries. The use of lignin offers substantial cost saving in manufacturing of carbon fiber. Use of lignin makes possible to produce carbon fiber based on renewable raw materials (jmrt.com.br).

With a sustainable appeal, use of lignin presents many benefits in comparison with the current commercial precursors (Table 1.8).

Table 1.8 Advantages of lignin over other precursors like MPP and PAN.

Very inexpensive
A renewable product
Already substantially oxidized so that it can be oxidatively thermostabilized at potentially much higher rates than either MPP or PAN

These benefits are limited not only to the environment-friendly approach but also to production advantages and a rentable manufacturing process. Production of carbon fiber from lignin is less-energy intensive and releases about 22% less carbon dioxide equivalent greenhouse gas emission in comparison with conventional polyacrylonitrile-based carbon fiber. Also the production is quite faster, and environmental toxicity is lower. According to ORNL the finished carbon fiber production cost reduces by about 30%–40% (iopscience.iop.org; Das, 2011).

References

Aoki K, Usui Y, Narita N, Ogiwara N, Iashigaki N, Nakamura K, Kato H, Sano K, Ogiwara N, Kametani K, Kim C, Taruta S, Kim YA, Endo M, Saito N. A thin carbon-fibre web as a scaffold for bone-tissue regeneration. Small 2009;5:1540–6.

Bahl OP, Shen Z, Lavin JG, Ross RA. Manufacture of carbon fibres. In: Donnet J-B, Wang TK, Peng JCM, Reboulliat S, editors. Carbon fibres. New York: Marcel Dekker; 1998. p. 1–83.

Bajpai P. Carbon fibre from lignin. Springer briefs in material science, Springer (Springer Nature); 2017.

Barnes CP, Sell SA, Boland ED, Simpson DG, Bowlin GL. Nanofibre technology: designing the next generation of tissue engineering scaffolds. Adv Drug Deliv Rev 2007;59:1413–33.

Baughman RH, Zakhidov AA, De Heer WA. Carbon nanotubes—the route toward applications. Science 2002;297(5582):787–92.

Carson EG. The future of carbon fibre to 2017, global market forecast. Smithers Apex; 2012.

Chen MCW. Commercial viability analysis of lignin based carbon fibre [Master thesis]. Simon Fraser University; 2014.

Chung DL. Carbon fibre composites. 1994: Boston, MA: Butterworth-Heinemann; 19943–11.

Das S. Life cycle assessment of carbon fiber-reinforced polymer composites. Int J Life Cycle Assess 2011;16:268–82.

Donnet JB, Bansal RC. Carbon fibres. 2nd ed. New York: Marcel Dekker; 1990. p. 1–145.

Figueiredo JL, Bernardo CA, Baker RTK, Hüttinger KJ, editors. Carbon fibres filaments and composites. Dordrecht: Kluwer Academic; 1990. p. 3–4.

Fitzer E, Heine M. Carbon fibers in book: Industrial carbon and graphite materials—Raw materials, production and applications [in print]. Wiley-VCH; 2019.

Fitzer E, Edie DD, Johnson DJ. in: Figueiredo JL, Bernardo CA, Baker RTK, Huttinger KJ, editors. Carbon fibres filaments and composites. 1st ed. New York: Springer; 1989. p. 3–41. 43–72, 119–146.

Forrest A, Pierce J, Jones W, Hwu E. Low-cost carbon fibres. Synergy; 2001. Retrieved 18 August, from, http://virtual.clemson.edu/caah/synergy/ISSUE-1LCCF.htm.

Fuchs ERH, Field FR, Roth R, Kirchain RE. Strategic materials selection in the automobile body: economic opportunities for polymer composite design. Compos Sci Technol 2008;68:1989–2002.

Gellerstedt G, Tomani P, Axegard P, Backlund B. Lignin recovery and lignin based products. In: Christopher L, editor. Integrated forest biorefineries: challenges and opportunities. Cambridge, UK: Royal Society of Chemistry; 2012. p. 1–66.

Hajduk F. Carbon fibres overview. In: Global outlook for carbon fibres 2005. Intertech conferences. San Diego, CA, 11–13 October 2005; 2005.

Hedge R, Dahiya A, Kamath MG. Materials science and engineering, Available from: http://web.utk.edu/~mse/Textiles/CARBON%20FIBERS.htm; 2004.

Holmes M. Global carbon fibre market remains on upward trend. Reinf Plast 2014;58:38–45.

Huang X. Fabrication and properties of carbon fibres. Materials 2009;2:2369–403. https://doi.org/10.3390/ma2042369.

Inwood J. Sulfonation of Kraft lignin to water soluble value added products [Master's thesis]. Thunder Bay, ON: Lakehead University; 2014.

Minus ML, Kumar S. The processing, properties and structure of carbon fibres. JOM 2005;57(2):52–8.

Minus ML, Kumar S. Carbon fibre. Kirk-Othmer Encycl Chem Technol 2007;26:729–49.

Mohan D, Pittman Jr CU, Steele PH. Single, binary and multi-component adsorption of copper and cadmium from aqueous solutions on Kraft lignin—a biosorbent. J Colloid Interface Sci 2006;297(2):489–504. https://doi.org/10.1016/j.jcis.2005.11.023.

Nolan L. Carbon fibre prostheses and running in amputees: a review. J Foot Ankle Surg 2008;14:125–9.

Norgren M, Edlund H. Lignin: recent advances and emerging applications. Curr Opin Colloid Interface Sci 2014;19(5):409–16. https://doi.org/10.1016/j.cocis.2014.08.004.

Ogawa H. Architectural application of carbon fibres: development of new carbon fibre reinforced glulam. Carbon 2000;38:211–26.

Olenic L, Mihailescu G, Puneanu S, Lupu D, Biris AR, Margineanu P, Garabagiu S. Investigation of carbon nanofibres as support for bioactive substances. J Mater Sci Mater Med 2009;0957-4530. 20(1):177–83.

Pimenta S, Pinho ST. Recycling carbon fibre reinforced polymers for structural applications: technology review and market outlook. Waste Manag 2011;31:378–92.

Red C. Aerospace will continue to lead advanced composites market in 2006. Compos Manuf 2006;7:24–33.

Roberts T. The carbon fibre industry: global strategic market evaluation 2006–2010. Watford, UK: Materials Technology Publications; 2006. p. 10 93–177, 237.

Saito N, Aoki K, Usui Y, Shimizu M, Hara K, Narita N, Ogihara N, Nakamura K, Ishigaki N, Kato H, Haniu H, Taruta S, Kim YA, Endo M. Application of carbon fibres to biomaterials: a new era of nano-level control of carbon fibres after 30-years of development. Chem Soc Rev 2011;40(7):3824–34.

Soutis C. Fibre reinforced composites in aircraft construction. Prog Aerosp Sci 2005;41:143–51.

Souto F, Calado V, Pereira Jr N. Lignin-based carbon fiber: a current overview. Mater Res Express 2018;5. 072001. https://doi.org/10.1088/2053-1591/aaba00.

Thostensona ET, Renb Z, Choua TW. Advances in the science and technology of carbon nanotubes and their composites: a review. Compos Sci Technol 2001;61:1899–912.

Todd J. How is carbon fiber made? ThoughtCo; 2019. April 21, 2019, thoughtco.com/how-is-carbon-fiber-made-820391.

Tran PA, Zhang L, Webster TJ. Carbon nanofibres and carbonnanotubes in regenerative medicine. Adv Drug Deliv Rev 2009;61(12):1097–114. https://doi.org/10.1016/j.addr.2009.07.010.

van der Woude HV, de Groot S, Janssen TWJ. Manual wheelchairs: research and innovation in rehabilitation, sports, daily life and health. Med Eng Phys 2006;2006(28):905–15.

Warren CD, Paulauskas FL, Baker FS, Eberle CC, Naskar A. Development of commodity grade, lower cost carbon fibre—commercial applications. SAMPE J 2009;45(2):24–36.

Watt W. In: Kelly A, Rabotnov YN, editors. Handbook of composites., vol. I. Holland: Elsevier Science; 1985. p. 327–87.

Witten E, Kraus T, Kühnel M. Composites market report. Industrievereinigung Verstärkte Kunststoffe; 2015.

Witten E, Kraus T, Kühnel M. Composite market report. Industrievereinigung Verstärkte Kunstsoffe; 2016.

Witten E, Sauer M, Kühnel M. Composites market report. Industrievereinigung Verstärkte Kunststoffe; 2017.

Xiaosong H. Fabrication and properties of carbon fibers. Materials 2009;2:2369–403.

Zhang X, Shen Z. Carbon fibre paper for fuel cell electrode, Fuel 2002;81:2199–201. www.eere.energy.gov.

Further reading

Market Watch. Toray to boost carbon fibre production, http://www.marketwatch.com/story/toray-to-boost-carbon-fibre-production-2012-03-08.
Specialty Fabrics Review. Cars will drive up demand for carbon fibres, http://specialtyfabricsreview.com/articles/0910_sw14_carbon_fibres.html.

CHAPTER 2

Raw materials and processes for the production of carbon fiber

The raw material used for making carbon fiber is termed precursor. Polyacrylonitrile (Fig. 2.1A) is presently the main precursor for producing carbon fibers because of the combination of tensile and compressive properties and also the carbon yield (Chung, 1994; Watt, 1985; Donnet and Bansal, 1990; Minus and Kumar, 2005, 2007a,b; Fitzer, 1990; Xiaosong, 2009; Peebles, 1995; Ozcan et al., 2014). About 90% of the carbon fibers are made from polyacrylonitrile. DuPont in the 1940s developed these fibers for use as textile fiber. The remaining 10% are made from pitch (Fig. 2.1B) (Donnet and Bansal, 1990; Minus and Kumar, 2005, 2007a,b). These materials are organic polymers, having long strings of molecules linked by carbon atoms. The exact composition of each precursor is found to vary from one company to another and is usually a manufacturer's secret.

Different types of gases and liquids are used during the production process (Donnet and Bansal, 1990; Minus and Kumar, 2005). Some of these materials are designed in such a way that they react with the fiber for achieving a desired effect, whereas other materials are designed to avoid certain reactions with the fiber or not to react. The exact compositions of these process materials are considered manufacturer's secret as in case of precursors (Fig. 2.2).

2.1 Polyacrylonitrile based carbon fibers

Extensive research has been done on the production of polyacrylonitrile-based carbon fibers, and several patents have been obtained or filed (Chung, 1994; Donnet and Bansal, 1990; Minus and Kumar, 2005; Minus et al., 2006; Masahiro et al., 1984; Knudsen, 1963; Daumit et al., 1990a,b,c; Grove et al., 1988; Mukundan et al., 2006; Bajaj et al., 1998; Shiromoto et al., 1990; Masaki et al., 1998; Houtz, 1950; Schurz, 1958; Standage and Matkowshi, 1971; Goodhew et al., 1975; Clarke and Bailey, 1973; Bailey and Clarke, 1970; Henrici-Olive and Olive, 1979, 1983; Rangarajan et al., 2002; Ganster et al., 1991; Gupta and Harrison, 1996; Fitzer and Muller, 1975; Fitzer et al., 1986; Yoshinori et al., 1978; Hamada et al., 2001; Potter and Scott, 1972; Raskovic and Marinkovic, 1975, 1978; Deurberque and Oberlin, 1991; Turner and Johnson, 1973; Gump and Stuetz, 1977; Kishimoto and Okazaki, 1977a,b; Riggs,

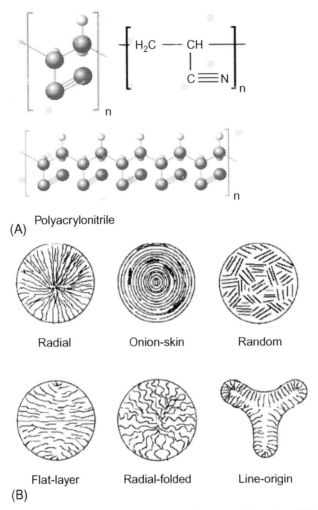

Fig. 2.1 Chemical structure of (A) polyacrylonitrile (www.shutterstock.com) and (B) microstructure of MP carbon fibers. *(Reproduced with permission Edie DD. The effect of processing on the structure and properties of carbon fibers. Carbon 1998; 36: 345–362.)*

1972a,b,c; Ko et al., 1988; Shiedlin et al., 1985; Mccabe, 1987; Mladenov and Lyubekeva, 1983; White et al., 2006; Karacan and Erdoğan, 2012; Paiva et al., 2003 Imai et al., 1990; Ohsaki et al., 1990; Hirotaka and Hiroaki, 2006; Kuwahara et al., 2008; Hunter, 1967; Warren et al., 2008; Dasarathy et al., 2002; Paulauskas et al., 2006; Sung et al., 2002; Smith et al., 1966; Barr et al., 1976; Otani, 1965; Otani et al., 1966; Otani, 1982; Singer, 1977; Lewis, 1977; Chwastiak and Lewis, 1978; Fu and Katz, 1991; Peter et al., 1988; Bolanos et al., 1993; Diefendorf and Riggs, 1980;

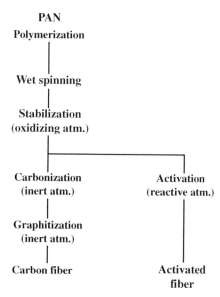

Fig. 2.2 Process for manufacture of carbon fiber from polyacrylonitrile. *Based on Chung DL. In Carbon fibre composites. Boston: Butterworth-Heinemann; 1994. pp. 3–11; Minus and Kumar, 2006.*

Angier and Barnum, 1980; Yamada et al., 1983; Otani, 1982; Kalback et al., 1993; Romine et al., 2004; Yamada et al., 1986; Seo et al., 1989, 1991; Morgon, 2005).

The process steps involved in the production of carbon fibers from polyacrylonitrile include the following (Donnet and Bansal, 1990; Minus and Kumar, 2005; Xiaosong, 2009):

- polymerization of polyacrylonitrile
- wet spinning of fibers
- thermal stabilization
- carbonization
- graphitization
- surface treatment.

Acrylonitrile is first mixed with methyl acrylate or methyl methacrylate and then treated in the presence of a catalyst in a conventional suspension or solution polymerization process to produce a polyacrylonitrile. It is then spun into fibers using one of many dissimilar methods. In some cases the plastic is mixed with certain type of chemicals and pumped through very small jets into a chemical bath or quench chamber. Here the plastic coagulates and gets solidified into fibers. This resembles with polyacrylic textile fiber production process. In other case the plastic mixture is heated and pumped through very small jets into a chamber where the solvents are evaporated and the solid fiber is left. In the spinning step the internal atomic structure of the fiber is produced. The fibers are washed and stretched for achieving the desired diameter. This step helps in aligning the molecules within the fiber and provides the basis for the formation of the tightly bonded carbon crystals after carbonization. Before the carbonization of the fibers, they should be chemically changed for converting their linear atomic bonding to a more thermally stable ladder bonding. This is performed by heating the fibers in air to approximately 200–300°C for 30–120 min. This causes the fibers to take oxygen from the air and rearrange their bonding pattern of the atoms. The stabilizing chemical reactions are complicated and involve many steps; some of which take place concurrently. They also produce their own heat, which should be controlled for avoiding overheating the fibers. The stabilization process uses different types of equipment and methods (Donnet and Bansal, 1990; Minus and Kumar, 2005; Xiaosong, 2009). In some cases the fibers are drawn through heated chambers arranged in series, whereas in other cases the fibers are passed on hot rollers and through beds of loose materials that are held in suspension by flowing of hot air. In some cases, heated air that is used is mixed with certain gases that chemically speed up the stabilization process. After the fibers are stabilized, they are heated to ~1000–3000°C temperature for several minutes in a furnace that is filled with a gas not containing any oxygen. Due to the absence of oxygen, the fibers are not burnt at elevated temperatures. The pressure of the gas inside the furnace is kept higher in comparison with the outside air pressure, and the points where the fibers enter and leave the furnace are sealed to remove oxygen away. As the fibers are heated, they start losing their noncarbon atoms, and a few carbon atoms, in the form of various gases including water vapor, ammonia, carbon monoxide, carbon dioxide, hydrogen, and nitrogen. As the noncarbon atoms are removed, the remaining carbon atoms produce tightly linked carbon crystals that are aligned parallel to the long axis of the fiber. In some cases, two furnaces operating at two different temperatures are used for controlling the heating rate during carbonization (Fitzer and Muller, 1975; Raskovic and Marinkovic, 1978).

> *After carbonization, the fibres have a surface that do not bond well with the epoxies and other materials used in composites. For giving the fibres better bonding, their surface is slightly oxidized. When the oxygen is added to the surface, better chemical bonding is achieved. It also etches and roughens the surface for obtaining better mechanical bonding. Oxidation is obtained by submerging the fibres in different gases such as air, carbon dioxide, or ozone; or in certain liquids such as sodium hypochlorite or nitric acid. The fibres can also be coated electrolytically by making the fibres the positive terminal in a bath filled with various electrically conductive materials. Carbon fibres used in composite are often coated or surface treated. This is done for improving interaction between the fibre surface and the matrix. Specific polar groups and/or roughness are developed by surface treatment. This results in development of on the surface for better interaction with the matrix.*
>
> **(Schurz, 1958; Potter and Scott, 1972; Xiaosong, 2009)**

Surface treatment can be oxidative or nonoxidative. Oxidative treatment is performed with oxygen, nitric acid, or other oxidizing agent. Nonoxidative treatment involves grafting of polymers. Vapor phase deposition of pyrolytic carbon on the surface of carbon fiber can also be done. Carbon fibers can be also treated with plasma and can be sized with epoxy resin or other polymers. This is performed for making them compatible with a certain type of matrix. Interlaminar shear strength of carbon fibers subjected to surface treatment is in the range of 30–90 MPa, whereas the Brunauer-Emmett-Teller (BET) surface area for these surface-treated carbon fibers is usually in the 25–60 m^2/g range. Carbon fibers get degraded in the presence of oxygen at temperatures beyond 400°C and are stable in inert environment up to above 2000°C; they can be protected from oxidative degradation by the use of a coating like silicon carbide, silicon nitride, boron nitride, and aluminum oxide.

The decomposition of polyacrylonitrile during production of carbon fiber is shown in Fig. 2.3.

Methods used for surface treatment of the carbon fiber were reviewed by Tiwari and Bijwe (2014). The physical, chemical, and morphological changes that occur in fiber properties were reported. These changes because of treatment resulted in improved

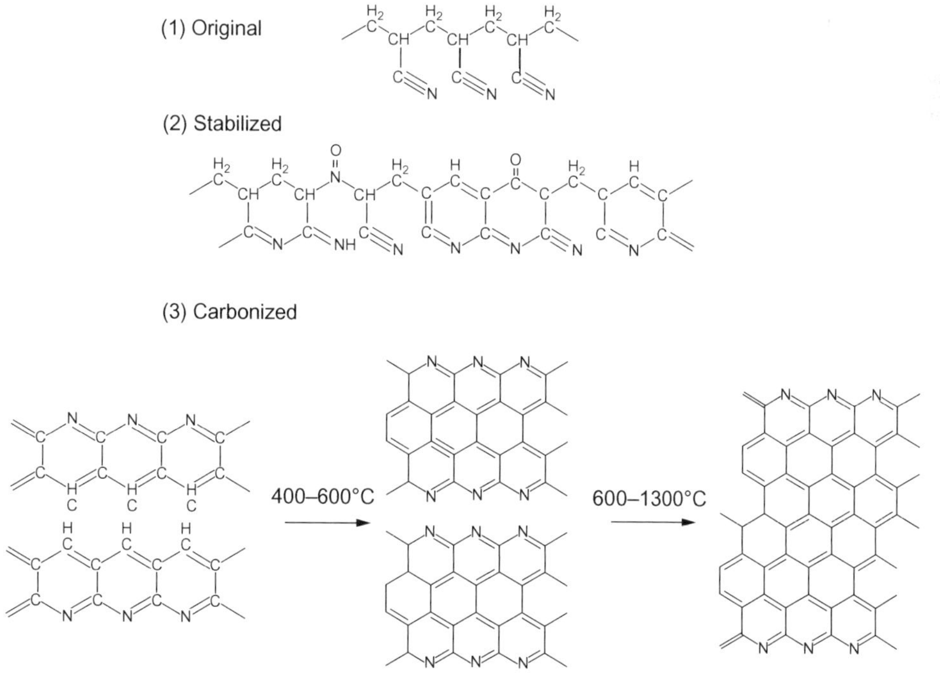

Fig. 2.3 Possible reactions occurring during the carbonization of stabilized polymer polyacrylonitrile precursor. *(Reproduced with permission from Zhu D, Xu C, Nakura N and Matsuo M. Study of carbon films from PAN/VGCF composites by gelation/crystallization from solution. Carbon 2002; 40(3): 363–373.)*

composite properties because of better surface area on fiber surface, chemical bonding, and adhesion between fiber and matrix. Treatment changed the morphology and increased the roughness of fiber surface. This increased surface area on fiber surface for improving interactions between fiber and matrix. Surface treatment also affects chemical structure of fibers and increases chemical bonding with matrix. Different treatments have different effects on fiber surface. Optimization is needed for selecting proper technique according to application and desired properties.

> *An aqueous guanidine carbonate treatment for increasing the thermal stabilization of Polyacrylonitrile precursor before carbonization was described. This chemical pretreatment proved that activity of guanidine carbonate is because of the presence of a carbon nitrogen double bond containing singly charged guanidinium cations in solution. Guanidinium ion is a strong base, having a pK value of 13.6, it promotes substantial resonance stabilization when protonated on the imine nitrogens.*
>
> *(Karacan and Erdoğan, 2012)*

The surface treatment process should be controlled properly for avoiding small surface defects, like pits, which may cause fiber failure. After this treatment the fibers are coated to protect them from damage during winding or weaving. This process is termed sizing. Coating materials are selected, which are compatible with the adhesive used for producing composite materials. The coating materials include epoxy, polyester, nylon, and urethane. The coated fibers are wound onto cylinders that are called bobbins. The bobbins are loaded into a spinning machine, and the fibers are twisted to form yarns of different sizes (Khosravani, 2012; b-composites.net).

Fig. 2.4 shows scanning electron micrographs of polyacrylonitrile-based carbon fibers (Kumar et al., 1993).

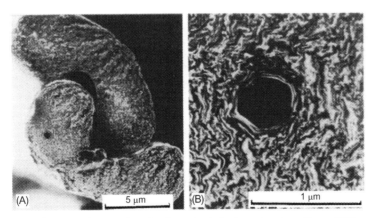

Fig. 2.4 Scanning electron micrographs of polyacrylonitrile-based carbon fibers: (A) low and (B) high magnification. *(Reproduced with permission Kumar S, Anderson DP and Crasto AS. Carbon fibre compressive strength and its dependence on structure and morphology, J Mater Sci 1993; 28: 423.)*

2.2 Pitch-based carbon fibers

Pitch is a viscoelastic material. It is made up of fused aromatic rings. Both isotropic and mesophase pitches are used for producing carbon fibers with low and high moduli. Pitch is manufactured from petroleum or coal tar. Mesophase pitch is produced by polymerizing isotropic pitch to a higher molecular weight (Mochida et al., 1990; Korai et al., 1991; Watanabe et al., 1999; Kofui et al., 1997; Kohn, 1976; Vakili et al., 2007; Yue et al., 2009; Sawaki et al., 1989; Sasaki and Sawaki, 1990; Miyajima et al., 2000; Bright and Singer, 1979; McHenry, 1977; Guigon et al., 1984a, 1984b; Singer, 1978; Riggs, 1985). Carbon fibers based on pitch show improved electrical and thermal properties in comparison with polyacrylonitrile-based fibers. Carbon fibers based on isotropic pitch were commercialized in the 1960s, but carbon fibers based on mesophase pitch were commercialized in 1980s (Donnet and Bansal, 1990; Minus and Kumar, 2005). Production of mesophase pitch–based carbon fiber is expensive in comparison with production from polyacrylonitrile (Fig. 2.5).

Production of carbon fibers from pitch involves the following steps (Fitzer, 1990; Chung, 1994; Watt, 1985; Donnet and Bansal, 1990; Minus and Kumar, 2005; Xiaosong, 2009; Mochida et al., 1990; Korai et al., 1991; Kohn, 1976; Yue et al., 2009; Sawaki et al., 1989):

- pitch preparation
- melt spinning

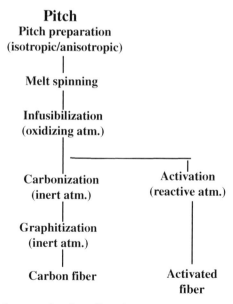

Fig. 2.5 Process for manufacture of carbon fiber from pitch. *Based on Chung DL. In Carbon fibre composites. Boston: Butterworth-Heinemann; 1994. pp. 3–11; Minus and Kumar, 2006.*

- thermosetting
- carbonization
- graphitization
- surface treatment

In pitch preparation process the major step is elimination of the impurities, which are solid particles, or gel-like materials. Presence of these impurities reduces the tensile strength of the product carbon fiber. Removal of low–molecular weight hydrocarbons is also very important. Pitch-based carbon fibers have modulus equal to graphite at 1050 GPa. This is considerably higher in comparison with the highest modulus achieved from polyacrylonitrile fibers of 650 GPa. The softening point of isotropic pitch is between 40°C and 120°C. The mesophase pitch is an anisotropic liquid crystal state of pitch containing aromatic molecules known as carbonaceous pitch. This has a softening point of about 300°C. It is produced by pyrolysis of isotropic pitch at a temperature range of 300°C and 500°C. Various methods are used for purifying isotropic and mesophase pitches before spinning. The molecular weight of pitch is usually in the range of 150–1000 g/mol with an average molecular weight of ~450 g/mol. Pitch is melt spun into a continuous fiber, which can be drawn. The spinning temperature for mesophase pitch is ~350°C. The cross section of the spinneret hole controls the shape of the fiber but can also be used for controlling the microstructure of the final carbon fibers. The transverse microstructures of the pitch-based carbon fibers also change with specific spinning conditions. Stabilization for pitch precursor fibers takes place at temperatures between 200°C and 300°C. Pitch fibers will soften and melt at higher temperatures as it behaves like a thermoplastic. The thermoplastic is converted to a thermoset during stabilization, and only then, it undergoes high-temperature carbonization. The degree of stabilization is controlled in a careful manner; otherwise, during carbonization, the fiber will melt if there is not enough stabilization. Prolonged stabilization results in the reduction in final carbon fiber mechanical properties. The stabilization time for isotropic pitch is usually higher as compared with the mesophase pitch. Pitch precursor fibers undergo carbonization and graphitization. Of all precursors the carbon yield for pitch-based fibers is the highest (70%–80%). Fig. 2.6 shows possible reaction mechanisms involved in the oxidative stabilization of pitch precursor, and Fig. 2.7 shows scanning electron micrographs of pitch-based carbon fibers: (a) low and (b) high magnification.

Pitch has several advantages compared with polyacrylonitrile. These are listed in the succeeding text (Fitzer, 1990; Chung, 1994; Watt, 1985; Donnet and Bansal, 1990; Minus and Kumar, 2005):

➢ reduced material cost
➢ higher char yield
➢ higher degree of orientation

The graphitic structure also gives pitch based carbon fibres higher elastic modulus and higher thermal and electrical conductivity along the fibre direction. However, the processing cost (mainly from pitch purification, mesophase formation and fibre spinning) to achieve high performance

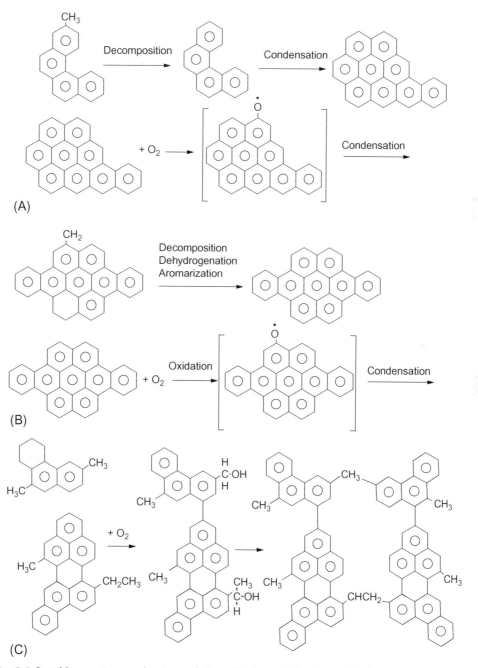

Fig. 2.6 Possible reaction mechanisms of the oxidative stabilization of pitch precursor. (a) NP80; (b) NHP; and (c) A60. *(Reproduced with permission Zeng SM, Maeda T, Tokumitsu K, Mondori J and Mochida I. Preparation of isotropic pitch precursors for general purpose carbon fibers (GPCF) by air blowing. II. Air blowing of coal tar, hydrogenated coal tar, and petroleum pitches. Carbon 1993; 31(3): 413–419.)*

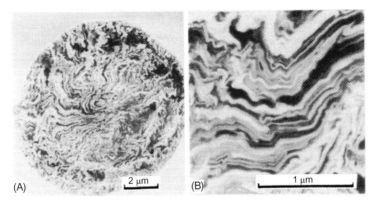

Fig. 2.7 Scanning electron micrographs of pitch-based carbon fibers: (A) low and (B) high magnification. *(Reproduced with permission Kumar S, Anderson DP and Crasto AS. Carbon fibre compressive strength and its dependence on structure and morphology, J Mater Sci 1993; 28: 423.)*

carbon fibres is higher. Pitch from petroleum and coal tar is isotropic. By evaporating low molecular weight fractions, isotropic pitch can be melt spun into low cost general purpose (low strength and low modulus) carbon fibres. To produce high performance fibres, an expensive hot stretching process needs to be applied. A more common way to produce high performance carbon fibres from pitch is to use an anisotropic pitch, such as mesophase pitch. Both isotropic and mesophase pitches are melt spinnable. Prior to fibre spinning, particulates are removed from the pitch. There is no need to hold the precursor fibres under a strong tension in the process of the stabilization and carbonization. The mesophase orients itself along the fibre axis direction during the precursor fibre spinning.

(www.mdpi.com)

2.3 Cellulose-based carbon fibers (CBCF)

Cellulose is the oldest known carbon fiber precursor material. Due to economic and ecological reasons, interest in producing biobased carbon fibers is increasing nowadays. Cellulose fibers undergo thermal decomposition without melting and have a crystalline structure. "It produces a strong carbonaceous material through pyrolysis. Cellulosic precursors have following properties:

➢ High thermal conductivity
➢ High purity
➢ Mechanical flexibility
➢ Low precursor cost

The regenerated cellulose fibers used for producing carbon fibers are textile-grade rayon, cuprammonium rayon or viscose rayon. The carbon fibers made from these precursors generally have some problems like large void content and inter filament de-bonding, which affect mechanical properties, resulting in weak and brittle carbon fibers, after

carbonization. Important parameters for the production of carbon fibers with good mechanical properties are listed below:
- Degree of polymerization
- Aspect ratio of individual nanofibrils within the hierarchical structure of cellulose
- The orientation of these nanofibrils along the fiber axis" (link.springer.com; Peng et al., 2003).

The effect of degree of polymerization of the cellulose on the mechanical properties was studied by Yoneshiga and Teranishi (1970). "Using cellulose with a degree of polymerization higher than 450 instead of conventional rayon (degree of polymerization ∼250), increased the tensile strength of the carbon fibres by about 30% after being carbonized at 800°C. Nonetheless, rayon fibre is still a main raw material for production of activated carbon fibres which are produced from carbon fibres using a high temperature process in an oxidizing atmosphere, resulting in a high surface area and a highly porous material" (Zeng et al., 2005).

Natural fibres, such as cotton and ramie, have not been preferred for carbonization due to their discontinuous filament structure, and low degree of orientation. They also contain impurities which arise from the complex structure of natural cellulose sources like lignin and hemicellulose. But, with development of new processing methods for making continuous fibres from natural cellulose fibres opens new possibilities to use full advantage of this exceptional structure for advanced applications. The development of the new generation regenerated cellulose (Lyocell) fibres has opened new possibilities for using cellulose for carbon-fibre production. Lyocell fibres are spun from cellulose solutions by dissolving cellulose pulp in N-methylmorpholine-N-oxide as the solvent. The benefit of using this solvent system is that polymer chains maintain a high degree of polymerization and Lyocell fibres have high crystallinity in which crystalline domains of cellulose-II structure are continuously oriented and dispersed along the fibre axis. These properties impart good strength to Lyocell. Lyocell also has desirable properties such as its round cross-section, and thermal stability. Therefore, Lyocell fibre is a promising precursor.

(Chanzy et al., 1990)

Several publications and patents support lyocell as a competitive precursor. But, its use for industrial carbon fiber production is not established so far. (Wu and Pan, 2002; Zhang et al., 2006; Peng et al., 2003; Ford and Mitchell, 1963; Richard et al., 1967; Yoneshiga and Teranishi, 1970; Zeng et al., 2005; McCorsley, 1981; Chanzy et al., 1990; Koslow, 2007; Wu et al., 2007).

For assuring a constant quality of the biobased precursor, man-made regenerated cellulose fibers, such as viscose or lyocell, are used. CBCFs were first produced for light bulb filament application (Fitzer, 1990; Chung, 1994. Watt, 1985; Donnet and Bansal, 1990; Minus and Kumar, 2005, 2007a,b; Xiaosong, 2009; Hajduk, 2005; Parry and Windle, 2012). Union Carbide in the 1960s carried out the commercial production of rayon-based carbon fibers for the first time. There are three major stages for rayon-based carbon fibers (Donnet and Bansal, 1990; Minus and Kumar, 2005):
- low-temperature decomposition
- carbonization
- graphitization

Rayon fibres are heated to 100°C in an inert atmosphere to remove water molecules. The temperature is gradually raised to 400°C; during which time, structural changes occur with a total weight loss of about 70%. The Rayon fibres are carbonized while stretching. The carbon yield for Rayon-based fibres range from 10% to 30%. The mechanical properties show improvement after graphitization with Young's modulus ranging from 170 to 500 GPa and tensile strength from 1 to 2 GPa for some commercial fibres. Rayon precursor is also being used for making activated carbon fibre. The production of rayon-based carbon fibres is now almost nonexistent.

(Minus and Kumar, 2005, 2007a,b)

One of the main challenges regarding CBCF is the low carbon yield compared with commercial polyacrylonitrile-based carbon fiber. The maximum theoretical carbon yield for a cellulose precursor is 44.4%. Without any pretreatment the actual yields in carbonization of cellulose are much lower, usually around 15%. Another challenge is the mechanical performance of CBCFs that is generally inferior to commercial polyacrylonitrile-based carbon fiber (Unterweger et al., 2018).

Currently, carbon fiber based on cellulose amounts only 1%–2% of the total production (Zhang et al., 2006). The major producers of CBCFs are.

(Dumanli and Windle, 2012) as follows:

- RUE-SPA-Khimvolokno (Republican Unitary Enterprise Svetlogorsk Production Association, Belarus)
- SGL Carbon, in Germany

Fig. 2.8 shows scanning electron micrographs of rayon-based carbon fibers. Fig. 2.9 shows scanning electron microscopy (SEM) micrographs of CBCFs (Ma et al., 2013).

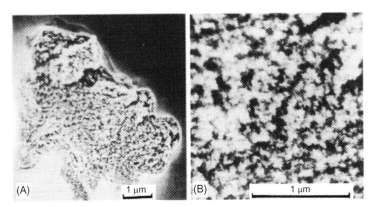

Fig. 2.8 Scanning electron micrographs of rayon-based carbon fibers: (A) low and (B) high magnification. *(Reproduced with permission from Kumar S, Anderson DP and Crasto AS. Carbon fibre compressive strength and its dependence on structure and morphology, J Mater Sci 1993; 28: 423.)*

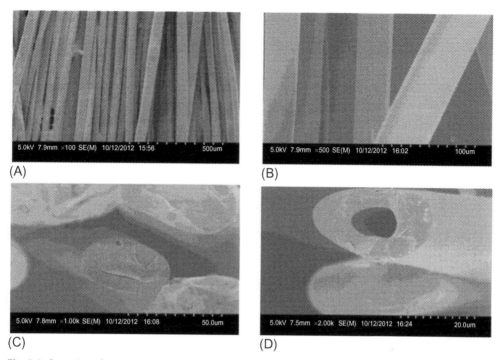

Fig. 2.9 Scanning electron microscopy (SEM) micrographs of cellulose-based carbon fibers (CBCFs): (A) and (B) side surface, (C) and (D) crossing section. *(Reproduced with permission Ma, X, Yuan, C, Liu, X. Mechanical, microstructure and surface characterizations of carbon fibers prepared from cellulose after liquefying and curing. Materials 2013; 7: 75–84.)*

2.4 Carbon fiber from lignin

Lignin-based carbon fiber has emerged as a material with high technological impact, economically attractive, and environmentally sustainable, although it is still far from being competitive with current precursors, showing signs of becoming a viable substitute for certain application areas, mainly those of the low-cost segment of the market, but with high demand.

(Souto et al., 2018)

In the lignin, high carbon content is present that makes it a promising candidate for the production of carbon fiber. But a suitable precursor should be able to form fiber by some of the spinning methods and tolerate the treatment conducted at high temperature. In lignins, thermoplasticity has been found that shows the possibility of using melt extrusion for fiber spinning, which is the preferred processing method as it involves less cost in comparison with wet spinning (Kadla et al., 2002a,b; Kubo and Kadla, 2005a,b,c).

Lignin shows many benefits over polyacrylonitrile and pitch for the production of commercial carbon fiber (Table 2.1).

Table 2.1 Advantages of extruding lignin rather than polyacrylonitrile.

Precursor is independent of fossil source
Increased productivity
Less energy expenditure in MJ per fiber kilogram
Minimal environment toxicity compared with other precursors
Reduced CO_2 per fiber kilogram compared with polyacrylonitrile
High carbon yield (up to 60%)

Based on Norgren M and Edlund H. Lignin: recent advances and emerging applications. Curr Opin Colloid Interface Sci 2014;19:409–416; Baker DA and Rials TG. Recent advances in low-cost carbon fibre manufacture from lignin. J Appl Polym Sci 2013;130:713–28; Luo J Lignin based carbon fibre [thesis master of science, chemical engineering] University of Maine USA. 2004; Luo J, Genco J, Cole B and Fort R. Lignin recovered from the near-neutral hemicellulose extraction process as a precursor for carbon fiber Bioresources 2011;6:4566–93; Chen MCW. Commercial viability analysis of lignin based carbon fibre [master dissertation], Burnaby, Canada: Simon Fraser University, 2014.

Use of lignin can promote a broader use of carbon fibers. "The main benefits of using lignin are its lower cost, high carbon content, high carbon yield during carbonization. It is also available in huge amounts. Furthermore, use of lignin helps in avoiding elimination products such as HCN or nitrous gases released during Polyacrylonitrile carbonization. But, unlike Polyacrylonitrile, lignin is a heterogeneous polymer, having different molecular weights, functional groups, and chemical linkages. This intrinsic heterogeneity is responsible for the inferior mechanical properties of current lignin based carbon fibers in comparison with carbon fibers manufactured from petroleum" (Xiaosong, 2009).

Oak Ridge National Laboratory in United States is active in pursuing research on lignin-based carbon fibre. Oak Ridge Carbon Fibre Composites Consortium was established in 2011 to accelerate the development of low-cost carbon fibre reinforced composite materials. The objectives set by the consortium is to manufacture a lignin-based carbon fibre with a tensile strength of 1.72 GPa and a modulus of 172 GPa that is suitable for the automotive industry. The consortium is having more than 52 members across the whole carbon fibre value chain starting from raw materials to downstream applications.

(Bajpai, 2017)

The major driving force behind the development of lignin carbon fiber is government regulatory changes on fuel consumption. In 2012 "The United States government legislated through updated Corporate Average Fuel Economy standards that the average fuel economy of cars and light trucks sold in the United States for model year 2017 will be 35.5 mpg, and will increase to 54.5 mpg for 2025 models. The most effective method for increasing fuel economy is to reduce the vehicle weight. Currently, Polyacrylonitrile based carbon fibre reinforced composites are able to offer up to 60% weight reduction at reduced cost (10 times). As more than 50% of the manufacturing cost is the cost of precursor, which fluctuates significantly with the price of oil. Department of Energy has invested well over $100 million over the last decade for examining possible routs of low-cost precursor alternatives. Lignin based precursor is one of the most promising contenders" (Berkowitz, 2011; Baker and Rials, 2013).

The development of an alternative precursor for carbon fiber, produced using renewable raw material, was identified many years ago. Kayocarbon fiber, from lignin, was manufactured by Nippon Kayaku Co. in Japan in 1960s. The mechanical properties of the carbon fiber were inferior, so this project was abandoned. The fiber was manufactured from lignosulfonate, using polyvinyl alcohol as the plasticizer, and then dry spun (Donnet and Bansal, 1990; Minus and Kumar, 2005). New attempts were made in the 1990s, and extensive work in his area was conducted (Sudo and Shimizu, 1987, 1989, 1992, 1994; Sudo et al., 1988; Yokoyama et al., 2003; Luo, 2004; Luo et al., 2011; Norberg, 2012; Norberg et al., 2012; Nordström, 2012; Nordstrom et al., 2012; Kubo et al., 1997, 1998; Kadla et al., 2002a,b; Baker, 2011; Kadla and Kubo, 2005; Kubo and Kadla, 2005a,b,c; Ito and Shigemoto, 1989; Uraki et al., 1995, 1997, 2001; Uraki and Kubo, 2006; Eckert and Abdullah, 2008). Dr. Baker's group has been quite active in his area (Baker, 2011; Baker, 2010a,b, 2011; Baker et al., 1969, 2005, 2008a, b, 2009a,b, 2010, 2011; Baker and Rials, 2013). Baker and Rials (2013) and Frank et al. (2012, 2014) published a review on this topic.

Several types of lignins—steam explosion lignin, organosolv lignin, and kraft lignin—have been studied as precursors for carbon fiber (Tang and Bacon, 1964; Peebles, 1994; Tang and Bacon, 1964; Bacon, 1974; Sudo and Shimizu, 1992; Yokoyama et al., 2003; Uraki et al., 1995; Kubo et al., 1998; Kadla et al., 2002a, b; Kubo and Kadla, 2005a,b,c; Braun et al., 2005,b; Luo, 2004; Fenner and Lephardt, 1981).

A process for the production of carbon fiber from hardwood lignin was developed. Lignin was produced using steam explosion technology from hardwood (birch). Hydrogenation process was used for modifying the lignin for improving melt spinning. Chloroform and carbon disulfide were used for dissolving and separating the insoluble lignin fraction. The lignin after purification was heated at temperatures between 300°C and 350°C under vacuum for 30 min. This produced viscous lignin having a softening point of 110°C and melted completely at 145°C. This material was found to be appropriate for the production of fine filaments. Infusible lignin fibers were produced by thermostabilizing the filaments at 210°C. The filaments were carbonized by heating from room temperature to 1000°C at a heating rate of 5°C/min in nitrogen (Yokoyama et al., 2003).

Sudo and Shimizu (1992) compared chemical structure of the precursor with the crude lignin. A substantial elimination of aliphatic functional groups in comparison with the starting material was found.

Phenolated lignin was also examined as a carbon fiber precursor. This study was done as an alternative to using hydrogenated lignin due to the high production cost of producing hydrogen. "Steam-exploded lignin was the feedstock used in the phenolysis process. The reaction was conducted by treating equal weights of phenol and crude lignin under

Table 2.2 Properties of carbon fiber from different types of lignin.

Lignin from different sources	Fiber diameter (μm)	Elongation (%)	Tensile strength (MPa)	Modulus of elasticity (GPa)
Hydrogenated hardwood lignin	7.6 ± 2.7	1.63 ± 0.29	660 ± 230	40.7 ± 6.3
Phenolated hardwood lignin	NA	1.4	455	32.5
Hardwood lignin from hardwood acetic acid pulping	14 ± 1.0	0.98 ± 0.25	355 ± 53	39.1 ± 13.3
Softwood lignin from hardwood acetic acid pulping	84 ± 15	0.71 ± 0.14	26.4 ± 3.1	3.59 ± 0.43
Hardwood Alcell lignin	31 ± 3	1.00 ± 0.23	388 ± 123	40.0 ± 14

Based on Sudo K and Shimizu K. A new carbon-fiber from lignin. J. Appl. Polym. Sci. 1992;44:127–134. https://doi.org/10.1002/app.1992.070440113; Yokoyama A, Nakashima N, Shimizu K. A new modification method of exploded lignin for the preparation of a carbon fiber precursor. J. Appl. Polym. Sci. 2003;48:1485–1491. https://doi.org/10.1002/app.1993.070480817; Kadla JF, Kubo S, Gilbert RD, Venditti RA, Compere AL and Griffith WL. Lignin-based carbon fibres for composite fibre applications. Carbon 2002a; 40(15):2913–2920, Kadla JF, Kubo S, Gilbert RD and Venditti RA. In: Hu TQ (ed.) Lignin-based carbon fibres, chemical modification, properties, and usage of lignin. New York: Kluwer Academic/Plenum, 2002b. pp 121–137.

vacuum at 180–300°C for 2–5 h. Para-toluene-sulfuric acid was used to catalyze the reaction. The resulting lignin pitch could be readily spun into fine filaments. These were converted into carbon fibre after thermo-stabilization and carbonization. The overall yield for the process was 43.7%" (Yokoyama et al., 2003; www.if.ufrrj.br).

Properties of the carbon fiber from phenolated lignin are shown in Table 2.2. The tensile strength of the carbon fiber was ~455 MPa. It was not as high as that of the hydrogenated lignin (www.if.ufrrj.br).

Softwood kraft lignin was acetylated for use in melt spinning lignin fiber (Schmidt et al., 1995). The acetylation process was performed with acetyl chloride, acetic anhydride, and acetic acid at temperatures of 70–100°C without catalyst. The acetylation reaction was also performed with a catalyst at reduced temperature. Organic amines were the preferred catalysts for producing lignin acetate that readily melted during spinning. The temperature for the reaction was ~50°C. The acetylation method allowed softwood lignin fiber to be spun at diameter between 5 and 100 μm. Data were not reported for the physical properties and overall carbon fiber yield (www.mdpi.com; www.if.ufrrj.br; Bajpai, 2017).

Organosolv lignin was used for producing carbon fiber. Acetic acid and ethanol were used as solvents to liberate the lignin. Crude lignin was obtained by acetic acid pulping of birch wood and used for the production of carbon fiber (Baker et al., 2005). The organosolv lignin was used without chemical modification outside of that done in the pulping process. The polydispersity of the lignin and partial acetylation of hydroxyl groups during pulping were actually responsible for the ability of the crude lignin to be readily spun into lignin fiber. Table 2.2 shows physical properties of carbon fiber produced from crude lignin obtained by acetic acid pulping of hardwood.

Kubo et al. (1998) used softwood lignin obtained by acetic acid pulping at atmospheric pressure for producing carbon fiber. Significant difference was found between Uraki et al. (1995) and Kubo et al. (1998) work. Uraki et al. (1995) used the crude lignin from hardwood obtained from the acetic acid pulping process. Kubo et al. (1998) removed the high molecular mass infusible fraction of the crude softwood lignin to allow melt spinning. The lignin was removed by filtering the crude lignin obtained from the acetic acid pulping process and then redissolved the low-molecular weight fractions in acetic acid at a lower concentration than used in the pulping process. This method dissolved the lower-molecular weight fractions of the crude lignin, which then reprecipitated. By this method, carbon fiber could be produced by direct carbonization, thereby eliminating the thermostabilization process. The physical properties of carbon fiber produced using this method are presented in Table 2.2.

> *In the Alcell process, a 50 wt% ethanol/water mixture was used to produce hardwood organosolv pulp, lignin and sugars. The Alcell pulping process was operated at temperatures between 190°C and 200°C, with corresponding operational pressure of 400–500 psig. The high operational pressure resulted from the high vapor pressure of ethanol. Lignin and sugar were recovered from displaced pulping liquor. Process water was used to precipitate the dissolved lignin which is recovered from the first-stage of spent liquor. The solid lignin was further purified by centrifugation, washed, dried and sold as a dried product. The product lignin can amount to approximately 18% by weight of the dry wood charge in the process. Alcell lignin is highly hydrophobic, low in ash and contains no sulfur and thus is distinctly different from either lignosulfonates obtained from the sulfite process or Kraft thiolignin. Properties of Alcell lignin, important to the production of carbon fibre, are its low contamination, low number-average molecular weight (1000 Da), low softening point (145°C), low glass transition temperature (100°C) and small median particle size (20–40 μm).*
>
> *(Bajpai, 2017)*

A comparison of Alcell lignin with hardwood lignin and Indulin AT (softwood kraft lignin) was made (Kadla et al., 2002a,b). The Alcell lignin was provided by Repap Enterprises in Newcastle, New Brunswick. Indulin AT could not be spun into lignin fiber due to charring before melting. Both the Alcell and hardwood lignin could be melt spun into lignin fiber, but the Alcell lignin had significantly lower spinning temperature, ~140°C for Alcell lignin in comparison with 200°C for hardwood lignin. Infusible lignin fiber could be produced during the thermostabilization process using Alcell lignin fiber when the heating was done below 12°C per h. The physical properties of carbon fiber produced from Alcell lignin are presented in Table 2.2.

Kadla et al. (2002a,b) and Baker et al. (2011) examined lignin-polyethylene oxide (PEO) blends for producing carbon fiber. Kraft hardwood lignin was used without chemical modification for producing carbon fibers by thermal spinning followed by carbonization. By adding PEO to the lignin, fiber spinning was possible. The blends of lignin and PEO could be converted into fiber by adding 3%–5% PEO. At dose higher than 5% PEO, fiber fusing took place during thermal stabilization. The properties of carbon fiber from lignin-PEO blends are presented in Table 2.3.

Table 2.3 Physical properties of carbon fibers from blends of lignin-PEO.

Source	Fiber diameter (μm)	Elongation (%)	Tensile strength (MPa)	Modulus of elasticity (GPa)
Hardwood lignin	46±8	1.12±0.22	422±80	40±11
Lignin-PEO (97-3)	34±4	0.92±0.21	448±70	51±13
Lignin-PEO (95-5)	46±3	1.06±0.14	396±47	38±5

Based on Kadla JF, Kubo S, Gilbert RD, Venditti RA, Compere AL and Griffith WL. Lignin-based carbon fibres for composite fibre applications. Carbon 2002a; 40(15):2913–2920, Kadla JF, Kubo S, Gilbert RD and Venditti RA. In: Hu TQ (ed.) Lignin-based carbon fibres, chemical modification, properties, and usage of lignin. New York: Kluwer Academic/Plenum, 2002b. pp. 121–137.

The use of PEO into the blend did not improve the physical properties of the carbon fiber but improved fiber spinning. Kubo et al. (1998) explored several lignin–synthetic polymer blends as precursors for producing carbon fiber. Fiber produced from unmodified hardwood kraft and organosolv lignin was found to be brittle and was not handled easily. The problem was solved by using lignin and synthetic polymers of poly(ethylene terephthalate) (PET), polyethylene oxide (PEO), and polypropylene (PP) blends for reducing the brittleness and improving the physical properties of the spun fiber. Blends containing up to 25% of the synthetic polymer were produced.

"The physical properties of the lignin-based polymers were found to be dependent upon following three factors:
- The source and properties of the lignin
- The amount and physical properties of the synthetic polymer being incorporated
- Chemical interactions between the components. Addition of synthetic polymers into the precursor blends improve the ability of the lignin polymer blend to be spun into fiber, the brittleness of the spun fiber is reduced, and the flexibility is improved (www.if.ufrrj.br).

The physical properties of some lignin–polymer blends are summarized in Table 2.4.

Addition of poly(ethylene terephthalate) to lignin to form a precursor blend improved the modulus of elasticity and also the tensile strength of the carbon fiber (Table 2.4).

Table 2.4 Physical properties of carbon fibers from blends of lignin.

Source of lignin	Fiber diameter (μm)	Elongation (%)	Tensile strength (MPa)	Modulus of elasticity (GPa)
Lignin	NA	1.07	422	39.6
Lignin-PET (75–25)	NA	0.77	511	66.3
Lignin-PP (75–25)	NA	0.50	113	22.8

Based on Kubo S and Kadla JF. Kraft/lignin/poly(ethylene oxide) blends: effect of lignin structure on miscibility and hydrogen bonding. J Appl Polym Sci 2005b;98:1437–1444.

Physical properties were not improved by the addition of polypropylene to lignin-polymer blend precursor (Kubo et al., 1998).

Kadla et al. (2002b) studied the potential of using kraft lignin for manufacturing carbon fibers. Carbon fiber material, which is strong and lightweight, offers promise in many applications; however, the cost of raw materials, petroleum pitch, and polyacrylonitrile limits the demand for this material. For ascertaining the most suitable lignin for producing carbon fiber, the LignoBoost technique was used to isolate industrial kraft lignins from both softwoods and hardwoods, and these kraft lignins were then purified and characterized.

> *Ultrafiltration of the black liquor, prior to isolation, resulted in a kraft lignin of a satisfactorily high degree of purity. The type of kraft lignin used governed the lignin's response to thermal treatment. The lignins were rendered more stable by oxidative stabilisation, and there was a 10%–20% increase in the final yield after carbonisation, compared with stabilisation without oxygen. The products that were obtained suggested that radical, oxidation, condensation, and rearrangement reactions were the primary reactions occurring during oxidative stabilisation. Due to structural differences between kraft lignins from softwoods and from hardwoods, it was possible to carry out thermal stabilisation in an inert atmosphere using only heat for the softwood kraft lignin fibres. A one-step operation was all that was required to perform stabilisation and carbonisation on softwood kraft lignin fibres, suggesting that there is no need for a separate stabilisation step with these fibres, which may reduce processing costs.*
>
> *(Bajpai, 2017)*

Mechanical properties of carbon fibers made from lignin precursor are shown in Table 2.5.

Table 2.5 Mechanical properties of carbon fibers made from lignin.

Type of lignin	Fiber diameter, μm	Tensile strength, MPa	Tensile modulus, GPa	References
Ethanol lignin	31	388	40	Kadla et al. (2002a,b)
Acetic acid lignin, HW	35	155	15	Uraki et al. (1995)
Acetic acid lignin, SW	84	26	3.6	Kubo et al. (1998)
Steam explosion lignin	7.6	660	41	Sudo and Shimizu (1992)
Lignosulfonate	12–14	250	27	Fukuoka (1969)
Hardwood kraft	46	422	40	Kadla et al. (2002a,b)
Hardwood kraft + PEO	46	396	38	Kadla et al. (2002a,b)
Hardwood kraft	46	605	61	Kubo and Kadla (2005a,b,c)
Hardwood kraft + PET	31	669	84	Kubo and Kadla (2005a,b,c)

Based on Gellerstedt G, Sjöholm E and Brodin I. The wood-based biorefinery: a source of carbon fiber? Open Agric J 2010;3:119–124.

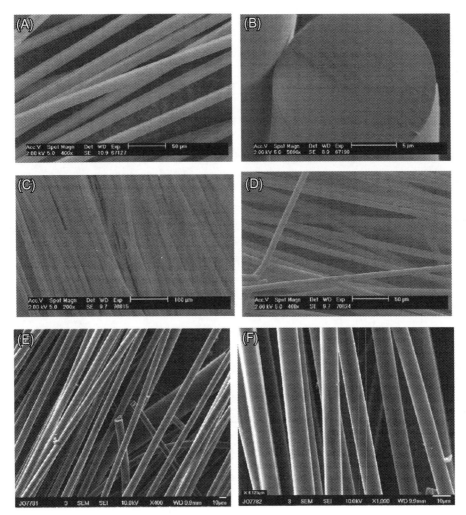

Fig. 2.10 Scanning electron microscopy images of (A and B) lignin fibers extruded from an organic solvent-extracted kraft hardwood lignin; (C and D) oxidized lignin fiber at heating rates of (C) 0.05°C min^{-1} and (D) 0.025°C min^{-1} showing different degrees of fusion; (E and F) carbon fiber, carbonized to 1000C at 2°C min^{-1} after stabilization at a heating rate of 0.01°C min^{-1}. *(Reproduced with permission Baker DA, Harper DP and Rials TG. Carbon fibre from extracted commercial softwood lignin. In: Book of abstracts of the fibre society 2012 fall conference, Boston: The Fibre Society, 2012a. pp. 17–18, Baker DA, Gallego, NC and Baker, FS. On the characterization and spinning of an organic-purified lignin toward the manufacture of low-cost carbon fiber. J Appl Polym Sci 2012b; 124: 227.)*

Scanning electron microscopy images of lignin fibers are shown in Fig. 2.10.

Lignin-based carbon fiber is the most value-added product from a wood-based biorefinery. The prospects of producing carbon fiber from kraft lignin are not clear, and there are several problems that should be solved. The laboratory produced carbon fiber from kraft lignin has strength properties and spinnability very much below those of commercial carbon fiber produced from polyacrylonitrile or pitch.

The knowledge about the structure of kraft lignin, the fractionation and purification of lignin, the thermal properties and the possibility of transforming lignin from thermoplastics to thermosetting polymers has substantially increased during the recent years. The limited commercial availability of kraft lignins will be changed by the measures taken by the forest industry and potentially by the automotive industry if they identify vehicle weight reduction as an option for future fuel efficient cars. In combination with a general interest to replace petrochemical raw materials with renewable raw materials, kraft lignin-based carbon fibre is promising but needs more research.

(www.benthamscience.com)

Many patents on the production of carbon fiber from lignin have been applied/issued in the last decade (Kadla et al., 2002a, 2002b; Baker, 2011; Kubo and Kadla, 2004; Kubo and Kadla, 2005a; Shen et al., 2007; Takanori et al., 2010; Bissett and Herriott, 2012; Sazanov et al., 2007; Seydibeyoglu, 2012; Lehmann et al., 2012a,b; Sevastyanova et al., 2010; Qin and Kadla, 2011, 2012; Scholze et al., 2001; Baker et al., 2011; Wohlmann et al., 2010; Shen et al., 2011; Maradur et al., 2012; Lehmann et al., 2012a,b; Uraki and Kubo, 2006; Ma and Zhao, 2010, 2011; Liu and Zhao, 2012; Lin et al., 2013; Kato et al., 2012; Lallave et al., 2007; Ruiz-Rosas et al., 2010; Hosseinaei and Baker, 2012a,b,c; Chatterjee et al., 2014).

Yang et al. (2011) produced carbon fibers from copolymers lignin based on lignin mixtures containing two or more of hardwood and softwood lignins and pitch.

Berlin (2011) produced lignin derivatives having a certain carbon and/or a certain alkoxy content and used for producing carbon fibers. Higher alkoxy and/or carbon contents improved the properties of carbon fibers.

Kim et al. (2011) manufactured lignin having high melt processability.

Wohlmann et al. (2010) developed a method for lignins having a Tg in the range 90–60°C, a polydispersity of <28, an ash and a volatiles content of <1% (wt).

Sjoholm et al. (2012) produced fractionated softwood and hardwood alkali lignins, their mixtures, and mixtures with original lignins and studied the extrusion properties of the materials.

Use of polypropylene instead of PEO was used in the production of hollow core carbon fibers and carbon fibers with high macroporosity but low mesoporosity (Kadla et al., 2002a,b; Baker, 2011; Kubo and Kadla, 2004; Kubo and Kadla, 2005a).

Blends of lignin and polyacrylonitrile and their thermal conversion chemistry have been reported (Bissett and Herriott, 2012; Sazanov et al., 2007; Sazanov et al., 2008; Seydibeyoglu, 2012; Lehmann et al., 2012a,b).

Recently, there has been an interest in the production of carbon fiber from lignin nanocomposite fibers.

Addition of modified montmorillonite organoclays into an organosolv lignin increased the efficacy of fiber spinning, and the strength of the lignin fiber increased twofold (Sevastyanova et al., 2010).

From biooil a pyrolytic lignin was obtained and thermally treated for increasing its molecular weight and glass transition temperature before compounding and fiber spinning (Qin and Kadla, 2011; Scholze et al., 2001). The clays were found to be well intercalated as shown by x-ray diffraction. Organoclays did not substantially improve the strength properties of the lignin fibers as found in case of organosolv lignin. The optimum dose was about 1 wt%.

Baker et al. (2011) examined carbon nanotubes for improving the thermal and electrical conductivity and strength, modulus of different lignins used for producing carbon fiber. With different lignins, organosolv lignin, purified hardwood kraft lignin, and a blend of softwood and the hardwood lignin, the carbon nanotubes could be added at a higher dose of 15 wt% before fibers could not be melt spun.

Shen et al. (2011) reported a method of producing activated carbon fibers using phenol-formaldehyde resin.

Maradur et al. (2012) produced copolymer using hardwood lignin and acrylonitrile with acrylonitrile/lignin ratio from 5:5 to 8:2. The lignin-polyacrylonitrile copolymer was dissolved in dimethyl sulfoxide and converted into fibers through wet spinning using a coagulation water bath.

The possibility of utilizing lignocellulosic and liquid wood precursors was explored (Lehmann et al., 2012a,b; Uraki and Kubo, 2006). These are not strictly lignin-based carbon fibers, but these are interesting material, as they could provide a route to low-cost carbon fiber by the use of low-value lignocellulosic streams, potential biorefinery waste streams, and/or sawdust.

Ma and Zhao (2011) mixed powdered wood with phenol and phosphoric acid at temperature of 160°C for 2.5 h. The mixture was then polymerized with hexamethylenetetramine at a dose of 5 wt% and heated to create cross-linking. The solution was then spun, and fiber stabilization was performed by immersing the fibers in a solution of formaldehyde and hydrochloric acid at 90°C for 2 h. The tensile strengths and moduli were 1.7 and 176 GPa, respectively (Ma and Zhao, 2010; Liu and Zhao, 2012). Similar syntheses were reported by Lin et al. (2013a) and Kato et al. (2012).

Lignin fibers were produced in a single-step process using electrospinning. Elemental composition, adsorption/desorption isotherms, and functional groups were studied as a function of conversion ordinate and platinum contents (Ruiz-Rosas et al., 2010).

Hosseinaei and Baker (2012a,b,c) produced nanofiber mats with smooth surfaces and without defects from a softwood kraft lignin feedstock purified by using solvent extraction before electrospinning.

A process for the conversion of Alcell hardwood and softwood lignins into carbon fibers was studied by Chatterjee et al. (2014). Lignin carbon fibers were produced from Alcell hardwood and softwood lignin precursors. Lignin fibers were produced by melt processing. The optimal extrusion parameters were found out for different lignin precursors.

Norgren and Edlund (2014) have reviewed emerging applications of lignin. Carbon nanofibers having diameters of 200 nm were successfully produced from lignin using electrospinning (Lallave et al., 2007). The structural alignment of the carbon fiber mats was found to be important in further improving the mechanical properties of the material (Lin et al. 2013). Stabilization and carbonization of kraft lignin fibers for producing carbon fiber can be conducted as a single-step process, just differing in temperature and treatment time. These findings are important in reducing the costs in commercialization of lignin as a precursor for carbon fibers (Norberg et al., 2013). The electrospinning of lignins for the production of fibrous networks has been also conducted (Dallmeyer et al., 2010). The mechanically flexible mats of carbon nanofibers produced from alkali lignin would be innovative and sustainable electrode materials for flexible high-performance supercapacitors (Lai et al., 2002). Saha et al. (2014) synthesized mesoporous carbon from lignin gels using structure-dictating agents for controlling porosity in a supercapacitor electrode material.

Many other attempts were made (Luo et al., 2011; Norberg, 2012; Lin et al., 2012; Nordström et al., 2013; Sudo and Shimizu, 1987; Uraki et al., 1995, 2001).

Each one added new insights to lignin melt spinning: (a) low molecular weight chains act as plasticizer for melt spinning while higher molecular weight chains provide desired viscosity and Tg; (b) neat lignin from conventional extraction, without any purification procedure, provides brittle fibers; (c) devolatilization procedure before melt spinning favors extrusion. It was due to this research that Baker and co-workers could obtain the correct specifications now known for lignin use as a carbon fiber precursor.

(Paul et al., 2015; iopscience.iop.org)

2.4.1 Problems with lignin

The softening temperature of lignin is in the range of 160–190°C. As a result, successive heating and carbonization of a lignin fiber will cause self fusion unless a stabilization reaction such as oxidation with oxygen (air), is included. Homogeneous oxidation throughout the whole cross-section of the lignin fiber needs a low rate of heating for obtaining oxidation rather than cross-coupling, dehydration and other competing reactions. To become industrially feasible the rate of modifying the lignin fiber into a thermo-setting material must be strongly increased before carbonization.
(Kubo et al., 1996, 1997, 1998; Brodin et al., 2009; Braun et al., 2005; www.benthamscience.com)

One or more purification steps before spinning are needed for using lignin as a precursor for carbon fiber. A set of preliminary purity data were obtained by Oak Ridge National Laboratory in the United States:
- <5 wt% volatiles measured at 250°C
- <1000 ppm inorganics (ash)
- <500 ppm nonmelting particulates >1 µm.

In earlier attempts to produce carbon fiber from lignin, through purification of the lignin was conducted before spinning. Low-molecular weight products were removed by treating lignin in vacuum at high temperature, and inorganics were reduced through extensive washing with aqueous acid. Organic solvents were also used (Kadla et al., 2002a,b; Uraki et al., 1995; Kubo et al., 1997; Kubo and Kadla, 2005a,b,c).

Only hardwood lignins have successfully resulted in carbon fiber. Softwood kraft lignin can be used as a precursor if a highly purified hardwood kraft lignin fraction, obtained by using solvent extraction, was added as a plasticizer (Baker et al., 2009a,b). As softwood kraft pulping is available in abundance in the United States, the potential availability of softwood kraft lignin is high.

Polydispersity of softwood kraft lignin is higher in comparison with that of hardwood. For unpurified softwood lignin, polydispersity values are in the range of 4.5 and for hardwood lignin it is 3–4 (Brodin et al., 2009; Glasser et al., 1993; Mörck et al., 1986, 1988).

> *By solvent fractionation, ~70% of birch kraft lignin and 50%–60% of softwood kraft lignin can be isolated as highly homogeneous polymeric material with polydispersity <2 and a low weight average molecular mass. Fractionation of the black liquor with a ceramic membrane followed by precipitation with carbon dioxide and acid according to the LignoBoost principle can be used as an alternative). Much more homogeneous lignin fractions were obtained in which carbohydrate impurities were virtually removed and the differences in polydispersity between softwood and hardwood kraft lignins were insignificant.*
>
> **(Mörck et al., 1986, 1988; Brodin et al., 2009; www.benthamscience.com)**

Heating kraft lignin to temperatures about 250°C resulted in a material loss of approximately 10%–15%. Small amounts of very volatile phenols, such as guaiacol, are lost, but removal of water appears to be the main reaction (Brodin et al., 2010). Further, XPS analysis of kraft lignin in air at 280°C or lower did not show an increase of olefin linkages (Braun et al., 2005). Kraft lignin precipitated from aqueous solution may contain considerable amounts of water resulting in bridging of individual lignin molecules through hydrogen linkages. Purified hardwood lignins produced using an aqueous isolation process are not melt-spinnable without addition of a plasticizing agent. Treatment of lignin at high temperature in the presence of water resulted in char formation as a main reaction as water acts as a strong reducing agent under these conditions. If a similar mechanism is operating under the conditions of lignin fiber stabilization, the presence of water may affect the chemical modification reactions negatively (www.benthamscience.com; Baker et al., 2009a,b; Kleinert and Barth, 2007).

2.5 Gas-phase grown carbon fibers

These are produced by decomposing gaseous hydrocarbons at temperatures between 300°C and 2500°C in the presence of Fe or Ni catalyst that is fixed to a substrate or fluidized in space (Donnet and Bansal, 1990; Minus and Kumar, 2005). Carbon, silicon, and quartz are the typical substrates, whereas hydrocarbons can be benzene, acetylene, or natural gas. Many reports are available on the development of gas phase-grown carbon fibers (Endo, 1988; Koyama, 1972; Anon, 1957; Tibbetts and Devour, 1986). Efforts were made to commercialize these fibers in the 1950s by Pittsburgh Coke and Chemical Company, but success was not achieved. The properties of the carbon fibers are affected by the temperature of the furnace and residence time of thermal decomposition. Diameters of these fibers ranged from 0.1 to 100 mm with circular, helical, and twisted cross sections.

Gas phase-grown carbon fibers contain circular cross sections and central hollow cores having diameters of few nanometer. The graphite networks are organized in concentric cylinders, and carbon fibers are placed like tree rings for forming a special structure having remarkable physical properties. These fibers have following properties:
- high tensile strength
- tensile modulus
- low electrical resistivity
- high thermal conductivity.

These properties make these fibers suitable for industrial applications. Their thermal conductivity (1950 W/m K) is the highest among all commercially available carbon fibers. These fibers deliver polymer composites having multifunctional properties. The main factors driving the gas phase-grown carbon fiber market worldwide are
- fast development of electronic devices,
- growth in the automotive sector,
- increase in the demand for high-performance products in aerospace and defense.

But, high price and long production cycles may check the market growth. In spite of that the increasing use of gas phase-grown carbon fiber is being increasingly used in the medical industry, and technological developments will provide opportunities for market expansion (https://www.alliedmarketresearch.com/gas-phase-grown-carbon-fiber-market).

2.6 Carbon nanotubes (CNTs)

A carbon nanotube made of carbon is a tube-shaped material having a diameter measuring on the nanometer scale (Donnet and Bansal, 1990; Minus and Kumar, 2005). "A nanometer is one-billionth of a meter, or about one ten-thousandth of the thickness of a human hair. The graphite layer looks like a rolled-up chicken wire having a continuous unbroken hexagonal mesh and carbon molecules at the apexes of the hexagons. Carbon Nanotubes have several structures, which differ in length, thickness, and in the type of helicity and number of layers. Although they are produced from the same graphite sheet, their electrical characteristics are different depending on these variations, acting either as metals or as semiconductors. As a group, Carbon Nanotubes usually have diameters ranging from <1 nm up to 50 nm. Their lengths are usually several microns, but the advancements have made the nanotubes much longer, and measured in centimeters" (www.nanocyl.com; Endo et al., 1976; Treacy et al., 1996; Thess et al., 1996; Dresselhaus and Eklund, 2000; Dresselhaus et al., 1995).

Carbon nanotubes can be classified into
- Single-wall nanotubes: Single-wall nanotubes contain a single graphene layer rolled into a seamless cylinder and can be semiconducting or metallic depending on the diameter and chiral angle of the tube. The typical diameter is 0.7–1.5 nm.
- Double-wall nanotubes: Double-wall nanotubes contain two concentric tubes. The typical diameter is 2–5 nm.
- Multiwall nanotubes: Multiwall nanotubes are composed of more than two concentric tubes. The typical diameter is 5–50 nm.

Nanotubes are synthesized by the following processes (Donnet and Bansal, 1990; Minus and Kumar, 2005; www.nanocyl.com):
- discharge
- catalytic chemical vapor deposition
- the high-pressure carbon monoxide processes

For more details on carbon nanotubes, please see Appendix.

2.7 Other carbon fiber precursors

Several other polymers have also been studied for their potential as raw materials for carbon fiber (Xiaosong, 2009; Iijima and Ichihashi, 1993; Majibur et al., 2007; Bengisu and Yilmaz, 2002; Prauchner et al., 2005; Sliva and Selley, 1975; Krutchen, 1974; Mavinkurve et al., 1995; Horikiri et al., 1978; Ashitaka et al., 1983, 1984; Newell and Edie, 1996; Kawamura and Jenkins, 1972; Jenkins and Kawamura, 1971; Yokota et al., 1975; Santangelo, 1970; Economy and Lin, 1976; Jiang et al., 1991; Murakami, 1989; Ezekial and Spain, 1967; Shindo et al., 1969; Soehngen and Willians, 1977).

Silk, chitosan, and eucalyptus tar pitch have been evaluated. These raw materials can reduce the production cost. Synthetic polymers have also been examined (Xiaosong, 2009). Linear and cyclic polymers have been also studied (Newell and Edie, 1996; Kawamura and Jenkins, 1972; Jenkins and Kawamura, 1971; Yokota et al., 1975; Santangelo, 1970; Economy and Lin, 1976; Jiang et al., 1991; Murakami, 1989; Ezekial and Spain, 1967; Shindo et al., 1969). These include
- phenolic polymers
- polyacenephthalene
- polyamide
- polyphenylene
- poly(p-phenylene benzobisthiazole)
- polybenzoxazole
- polybenzimidazole
- polyvinyl alcohol
- polyvinylidene chloride
- polystyrene.

Linear precursors need heat stretching for obtaining high-performance carbon fibers; their carbon yields are generally very low.

> US patent 4,070,446 disclosed a process of manufacturing carbon fibers from polyethylene. Melt spun polyethylene fibres were immersed in chlorosulfonic acid or sulfuric acid, fuming sulfuric acid or a mixture at 80°C for about 90 min for sulfonation. The washed and dried fibres were then carbonized in argon by increasing the temperature to 1200°C, under a tension of 16 mg/denier. The carbonization yield was 75%. The modulus and tensile strength were 139 and 2.5 GPa, respectively with the application of tension.
>
> *(www.mdpi.com; Horikiri et al., 1978)*

Sliva and Selley (1975) have patented a process for producing carbon fibers from polyacetylene. "Precursor fibres were solution spun and coagulated in a solvent bath containing acetone. The coagulated fibre tow was then stabilized at 260°C and 360°C, respectively. The carbonization was performed at 1100°C. The carbon fibres were further graphitized at a constant strain of 100% at 2500°C for about 2 min to high mechanical properties. The Young's modulus of carbon fibres was in the range of 350–490 GPa" (www.mdpi.com).

The production of carbon fibers from selectively polymerized poly(vinylacetylene) by the polymerization of vinyl groups in monovinylacetylene was studied (Mavinkurve et al., 1995). The polymer was melt spinnable. The precursor fibers were subjected to UV treatment (300 nm) for a few hours in a nitrogen atmosphere and then heated to 225°C in nitrogen. Oxidation was performed at 225°C in air. The polymer was melt spinnable. The pendent acetyl group was responsible for the cyclization reactions to form a conjugated ladder structure similar to the nitrile in polyacrylonitrile stabilization.

A melt extrudable mixture of a polyacetylene copolymer was studied. "Polyacetylene was produced by the polymerization of meta-diethynylbenzene and para-diethynylbenzene with dipropargyl ethers of dihydric phenols. The plasticizer used was dichlorobenzene or pyridine. The melt spun fibres were drawn for increased orientation and decreased diameter. Fibres with about 25 μm diameter could be stabilized at 310°C for 0.5 s and then at 200°C and 300°C for 20 s either in the presence or absence of oxygen. The thermally stabilized fibres were converted to carbon fibres at 1000°C and the resultant carbon fibres showed a tensile strength of about 2.3 GPa and a modulus of about 386 Gpa" (www.mdpi.com; Krutchen, 1974).

> Nagasaka et al. (1978) described a process for producing polybutadiene carbon fibers. The fibres were cyclized and then cross-linked in a solution of a Lewis acid in an inert organic liquid and treated in a solution of sulphur at 170–300°C for aromatization. The fibres were carbonized at 750–1500°C in nitrogen and graphitized at 1500–3000°C in argon. The carbon fibres carbonized without tension had a high carbonization yield of about 89%, but comparatively lower mechanical properties of about 58–71 GPa modulus and 0.95–1.2 GPa tensile strength.
>
> *(www.mdpi.com)*

Japanese researchers developed carbon fibers from syndiotactic 1,2 poly(butadiene) (s-PB). "The s-PB fibres were produced by melt spinning at temperature of 205°C. The precursor fibres were stabilized by immersing into a solution of Aluminum bromide in benzene at 42°C for 78 min under tension. The washed and dried fibres were then immersed into molten sulfur at 275°C for 14 min for dehydrogenation. The adhering sulfur was purged with nitrogen at 290°C for 7 min. Carbon fibres obtained by heating the precursors to 1400°C in an inert atmosphere showed a tensile strength of about 1.6 GPa and a modulus of about 139 GPa with a carbon yield of 82%. Carbon fibres obtained through heat treatments of up to 3000°C showed a tensile strength of about 2.0 GPa and a modulus of about 393 GPa with a carbon yield of 70%. The syndiotacticity induces the formation of thermally stable spiral ladder polymers" (www.mdpi.com; Ashitaka et al., 1983, 1984).

In another process, carbon fibers were obtained from poly(*p*-phenylene benzobisoxazole) that can be carbonized through a regular process into carbon fibers without stabilization. The resulting carbon fiber showed lower mechanical properties having a tensile strength of up to 1 GPa and a modulus of up to 245 GPa. The precursor fibers remained in the carbon fibers resulting in the low carbon fiber mechanical properties. The properties were improved by modification of the poly(*p*-phenylene benzobisoxazole) precursor fiber-spinning process.

The polymers with a high aromatic content usually offer a high carbon yield easy stabilization in some cases. But, these polymers do not produce high-performance carbon fibers and have high costs. There is a need to conduct research for reducing the processing cost while improving the mechanical properties of the resultant carbon fibers.

2.8 Safety concerns

The areas of concern in the production and handling of carbon fiber are dust inhalation, skin irritation, and the effect of fibers on electrical equipment. Additionally, the protective finish, or size, which is applied on the fiber may need additional safety precautions (Fitzer, 1990; Chung, 1994; Watt, 1985; Donnet and Bansal, 1990; Minus and Kumar, 2005; Minus and Kumar, 2007a,b). During processing, fine carbon elements break off and circulate in the air in the form of a fine dust. Health studies show that, unlike some asbestos fibers, carbon fiber is too large to be a health hazard if inhaled. But they can be an irritant, and people working in the area should wear protective masks. Smaller diameter carbon fiber can enter the respiratory tract, but no evidence of respiratory damage is found (Nagasaka et al., 1978; Owen et al., 1986; Jones et al., 1982; Thompson, 1989). They often create discomfort, but protective masks are recommended when in areas where carbon fiber dust is present.

Carbon fiber causes skin irritation, particularly on the back of hands and wrists (which are more sensitive). Protective clothing and barrier skin creams are recommended in areas where carbon fiber dust is present. The sizing materials used for coating the fibers often contain chemicals that may cause harsh skin reactions, so skin also needs protection. Carbon fiber are good conductors of electricity and can cause arcing and shorts in electrical equipment. "The best option is to locate sensitive equipment in clean rooms outside of areas where Carbon fiber is being processed. If this is not possible, electrical cabinets should be sealed effectively for preventing contact with Carbon fiber. A filtered air-positive purge provides additional protection for sensitive equipment" (Minus and Kumar, 2007a,b).

References

Angier DJ and Barnum HW (1980). Neomesophase formation. US Patent 4184942

Anon. Chem Eng October 1957;172–4.

Ashitaka H, Ishikawa H, Ueno H, Nagasaka A. Syndiotactic 1,2-polybutadiene with Co-CS$_2$ catalyst system. 1. Preparation, properties, and application of highly crystalline syndiotactic 1,2-polybutadiene. J Polym Sci A Polym Chem 1983;21:1853–60.

Ashitaka H, Kusuki Y, Yakamoto S, Onata Y, Nagasaki A. Preparation of carbon fibers from syndiotactic 1,2-polybutadiene. J Appl Polym Sci 1984;29:2763–76. https://doi.org/10.1002/app.1984. 070290907.

Bacon R. Carbon fibers from rayon precursors. In: Walker Jr. PL, Thrower PA, editors. Chemistry and physics of carbon. vol. 9. New York, NY, USA: Marcel Dekker; 1974. p. 1.

Bailey JE, Clarke AJ. Carbon fibres. Chem Br 1970;6:484–9.

Bajaj P, Paliwal DK, Gupta AK. Influence of metal ions on structure and properties of acrylic fibers. J Appl Polym Sci 1998;67:1647–59.

Bajpai P. Carbon fibre from lignin, Springer briefs in material science. Springer (Springer Nature, Switzerland AG.); 2017.

Baker FS. Low cost production of carbon fibre from sustainable resource materials: utilization for structural and energy efficiency applications. In: Presentation to Ontario Bio-Auto council, April 8–9, 2010. SEM images with higher definition courtesy of Paul Menchhofer of Oak Ridge National Laboratory. 2010.

Baker FS. Presentation at 2010 DOE hydrogen program and vehicle technologies annual merit review and peer evaluation meeting, June 7–11, 2010, Available from: http://www1.eere.energy.gov/vehiclesandfuels/pdfs/merit_review_2010/lightweight_materials/lm005_baker_2010_o.pdf; 2010.

Baker FS. Utilization of sustainable resources for materials for production of carbon fibre structural and energy efficiency applications, In: Nordic wood biorefinery conference, Stockholm, Sweden, March 22–24, 2011; 2011.

Baker DA, Rials TG. Recent advances in low-cost carbon fibre manufacture from lignin. J Appl Polym Sci 2013;130:713–28.

Baker DA, Gallego NC, Baker FS. On the characterization and spinning of an organicpurified lignin toward the manufacture of low-cost carbon fibre. J Appl Polym Sci 1969;124(1):227.

Baker FS, Griffith WL, Compere AL. Low-cost carbon fibres from renewable resources. Automotive light weight materials, FY 2005 process report. 187–96.

Baker DA, Gallego NC, Baker FS. Extended abstract, In: Book of abstracts of the fibre society 2008 fall conference, Industrial Materials Institute, Montreal, Canada, October 1–3; 2008 Available from: http://www.thefibresociety.org/Assets/Past_Meetings/PastMtgs_Home.html.

Baker DA, Baker FS, Gallego NC. Extended abstract, In: Book of abstracts of the fibre society 2009 fall conference, October 27–30, 2009Athens, GA, USA: University of Georgia; 2009a Available from: http://www.thefibresociety.org/Assets/Past_Meetings/PastMtgs_Home.html.

Baker FS, Gallego NC, Baker DA. Progress report for lightweighting materials, part 7.A. 2009, Available from: http://www1.eere.energy.gov/vehiclesandfuels/pdfs/lm_09/7_low-cost_carbon_fibre.pdf; 2009.

Baker DA, Gallego NC, Baker FS (2010) SAMPE'10 conference and exhibition, Seatle, WA, May 17–20, 2010.

Baker FS, Baker DA and Menchhofer PA (2011) Carbon nanotube enhanced precursor for carbon fibre production and method of making a CNT-enhanced continuous lignin fibre. US Patent application 2011285049. To be assigned to Oak Ridge Oak Ridge National Laboratory.

Barr JB, Chwastiak S, Didchenko R, Lewis IC, Lewis RT, Singer LS. High modulus carbon fibers from pitch precursors. Appl Polym Symp 1976;29:161–73.

Bengisu M, Yilmaz E. Oxidation and pyrolysis of chitosan as a route for carbon fiber derivation. Carbohydr Polym 2002;50:165–75.

Berkowitz J. The CAFE numbers game: making sense of the new fueleconomy regulations. Car and Driver, Retrieved from 23 June 2014. http://www.caranddriver.com/features/the-cafe-numbers-game-making-sense-of-the-newfuel-economy-regulations-feature; 2011.

Berlin A (2011) Carbon fibre compositions comprising lignin derivatives. WO 2011097721 A1, 2011 assigned to Lignol Innovations Ltd.

Bissett PJ, Herriott CW (2012) Lignin/polyacrylonitrile-containing dopes, fibres and production methods. US 20120003471 & WO 2012003070, 2012, Weyerhaeuser NR Company.

Bolanos G, Liu GZ, Hochgeschurtz T, Thies MC. Producing a carbon fiber precursor by supercritical fluid extraction. Fluid Phase Equilib 1993;82:303–10. https://doi.org/10.1016/0378-3812(93)87154-S.

Braun JL, Holtman KM, Kadla JF. Lignin-based carbon fibers: oxidative thermostabilization of Kraft lignin. Carbon 2005;43:385–94. https://doi.org/10.1016/j.carbon.2004.09.027.

Bright AA, Singer LS. Electronic and structural characteristic of carbon-fibers from mesophase pitch. Carbon 1979;17:59–69. https://doi.org/10.1016/0008-6223(79)90071-X.

Brodin I, Sjöholm E, Gellerstedt G. Kraft lignin as feedstock for chemical products: the effects of membrane filtration. Holzforschung 2009;63:290–7.

Brodin I, Sjöholm E, Gellerstedt G. The behavior of Kraft lignin during thermal treatment. J Anal Appl Pyrol 2010;87:70–7.

Chanzy H, Paillet M, Hagege R. Spinning of cellulose from N-methylmorpholine N-oxide in the presence of additives. Polymer 1990;31:400–5.

Chatterjee S, Jones EB, Clingenpeel AC, McKenna AM, Rios O, McNutt NW, Keffer DJ, Johs A. Conversion of lignin precursors to carbon fibres with nanoscale graphitic domains. ACS Sustain Chem Eng 2014;2014(2):2002–10.

Chung DL. Carbon fibre composites. Boston: Butterworth-Heinemann; 19943–11.

Chwastiak S, Lewis IC. Solubility of mesophase pitch. Carbon 1978;16:156–7.

Clarke AJ, Bailey JE. Oxidation of acrylic fibres for carbon fibre formation. Nature 1973;243:146–50.

Dallmeyer I, Ko F, Kadla JF. Electrospinning of technical lignins for the production of fibrous networks. J Wood Chem Technol 2010;30:315–29.

Dasarathy H, Schimpf WC, Burleson T, Smith SB, Herren CW, Frame AC, Heatherly PW. Low cost carbon fibre from chemically modified acrylics, In: Proceedings of the international SAMPE technical conference, Baltimore, MD, USA, May 2002; 2002.

Daumit GP, Ko YS, Slater CR, Venner JG and Young CC (1990a). Formation of melt-spun acrylic fibres possessing a highly uniform internal structure which are particularly suited for thermal conversion to quality carbon fibres. US Patent 4935180.

Daumit GP, Ko YS, Slater CR, Venner JG, Young CC and Zwick MM (1990b). Formation of melt-spun acrylic fibres which are well suited for thermal conversion to high strength carbon fibres. US Patent 4933128.

Daumit GP, Ko YS, Slater CR, Venner JG and Young CC (1990c). Formation of melt-spun acrylic fibres which are particularly suited for thermal conversion to high strength carbon fibres. US Patent 4921656.

Deurberque A, Oberlin A. Stabilization and carbonization of PAN-based carbon fibers as related to mechanical properties. Carbon 1991;29:621–8. https://doi.org/10.1016/0008-6223(91)90129-7.

Diefendorf RJ and Riggs DM (1980). Forming optically anisotropic pitches. US Patent 4208267.

Donnet JB, Bansal RC. Carbon fibres. 2nd ed. New York: Marcel Dekker; 19901–145.

Dresselhaus MS, Eklund PC. Phonons in carbon nanotubes. Adv Phys 2000;49:705–814.

Dresselhaus MS, Dresselhaus G, Saito R. Physics of carbon nanotubes. Carbon 1995;33(7):883–91. https://doi.org/10.1016/0008-6223(95)00017-8.

Dumanli A, Windle AH. Carbon fibres from cellulosic precursors: a review. J Mater Sci 2012;47 (10):4236–50. https://doi.org/10.1007/s10853-011-6081-8.

Eckert RC and Abdullah Z (2008). Carbon fibres from Kraft softwood lignin. US Patent US 0318043 A1, 25 December 2008.

Economy J, Lin RY. Adsorption characteristics of activated carbon fibers. Appl Polym Symp 1976;1976 (29):199–211.

Endo M. Large scale production, selective synthesis and applications of carbon NT by CCVD process. Chemtech 1988;18:568–76.

Endo M, Koyama T, Hishiyama Y. Structural improvement of carbon fibers prepared from benzene. Jpn J Appl Phys 1976;15:2073–6.

Ezekial HM, Spain RG. Preparation of graphite fibers from polymeric fibers. J Polym Sci 1967;19:249–65.

Fenner RA, Lephardt JO. Examination of the thermal-decomposition of Kraft pine lignin by Fourier-transform infrared evolved gas-analysis. J Agric Food Chem 1981;29:846–9. https://doi.org/10.1021/jf00106a042.

Fitzer E. Figueiredo JL, Bernardo CA, RTK B, Huttinger KJ, editors. Carbon fibres filaments and composites. Dordrecht: Kluwer Academic; 1990. p. 3–4 43–72, 119–146.

Fitzer E, Muller DJ. The influence of oxygen on the chemical reactions during stabilization of PAN as carbon fiber precursor. Carbon 1975;13:63–9.

Fitzer E, Frohs W, Heine M. Optimization of stabilization and carbonization treatment of PAN fibres and structural characterization of the resulting carbon fibres. Carbon 1986;24:387–95.

Ford CE and Mitchell CV (1963). UNION CARBIDE CORP fibrous graphite. US Patent 3107152.

Frank E, Hermanutz F, Buchmeiser MR. Carbon fibres: precursors, manufacturing, and properties. Macromol Mater Eng 2012;297:493–501.

Frank E, Steudle LM, Ingildeev D, Spörl JM, Buchmeiser MR. Carbon fibres: precursor systems, processing, structure, and properties. Angew Chem Int Ed 2014;53:2–39.

Fu TW and Katz M (1991), Process for making mesophase pitch. US Patent 4999099.

Fukuoka Y. Carbon fiber made from lignin (Kayacarbon). Jpn Chem Q 1969;5(3):63–6.

Ganster J, Fink HP, Zenke I. Chain conformation of polyacrylonitrile: a comparison of model scattering and radial distribution functions with experimental wide-angle X-ray scattering results. Polymer 1991;1991(32):1566–73.

Glasser WG, Dave V, Frazier CE. Molecular weight distribution of (semi-)commercial lignin derivatives. J Wood Chem Technol 1993;13:545–59.

Goodhew PJ, Clarke AJ, Bailey JE. A review of the fabrication and properties of carbon fibres. Mater Sci Eng 1975;17:3–30.

Grove D, Desai P, Abhiraman AS. Exploratory experiments in the conversion of plasticized melt spun. PAN-based precursors to carbon fibers. Carbon 1988;26:403–11.

Guigon M, Oberlin A, Desarmot G. Microtexture and structure of some high-modulus PAN-based carbon fibres. Fibre Sci Technol 1984a;20:177–98. https://doi.org/10.1016/0015-0568(84)90040-X.

Guigon M, Oberlin A, Desarmot G. Microtexture and structure of some high tensile strength, PAN-base carbon fibers. Fibre Sci Technol 1984b;20:55–72. https://doi.org/10.1016/0015-0568(84)90057-5.

Gump KH and Stuetz DE, Stabilization of acrylic fibers and films. US Patent 4004053, 1977.

Gupta A, Harrison IR. New aspects in the oxidative stabilization of PAN-based carbon fibers. Carbon 1996;34:1427–45.

Hajduk F. Carbon fibres overview, In: Global outlook for carbon fibres 2005, Intertech conferences. San Diego, CA, October 11–13, 2005; 2005.

Hamada M, Hosako Y, Yamada, T and Shimizu T (2001). Acrylonitrile-based precursor fiber for the formation of carbon fiber, process for preparing same, and carbon formed from same. US Patent 6326451.

Henrici-Olive G, Olive S. Molecular interactions and macroscopic properties of polyacrylnitrile and model substances. Adv Polym Sci 1979;32:123–52.

Henrici-Olive G, Olive S. Chemistry of carbon fiber formation from polyacrylonitrile. Adv Polym Sci 1983;51:1–60.

Hirotaka S and Hiroaki K (2006). Manufacturing process of isotactic copolymer for carbon fibre precursor. JP2006016482.

Horikiri S, Iseki J and Minobe M (1978). Process for production of carbon fibre. US Patent 4070446.

Hosseinaei O, Baker DA. Electrospun carbon nanofibres from kraft lignin, http://www.thefibresociety.org/httpdocs/Assets/Past_Meetings/BooksOfAbstracts/2012_Fall_Abstracts.pdf; 2012.

Hosseinaei O, Baker DA. Extended abstract in book of abstracts of the fibre society 2012 fall conference, Boston Convention and Exhibition Center, Boston, USA, November 7–9, 2012, Available from: http://www.thefibresociety.org/Assets/Past_Meetings/PastMtgs_Home.Htm; 2012.

Hosseinaei O, Baker DA. Electrospun carbon nanofibers from Kraft lignin, http://www.thefibersociety.org/httpdocs/Assets/Past_Meetings/BooksOfAbstracts/2012_Fall_Abstracts.pdf; 2012.

Houtz RC. "Orlon" acrylic fiber: chemistry and properties. Text Res J 1950;20:786–801. https://doi.org/10.1177/004051755002001107.

Hunter WL. Lability of the α-hydrogen in polyacrylonitrile. J Polym Sci B Polym Phys 1967;5(1):23–6.

Iijima S, Ichihashi T. Nature 1993;363:6430–603.

Imai K., Sumoto M., Nakamura H and Miyahara N (1990). Process for preparing carbon fibers of high strength. US Patent 4902762.

Ito K and Shigemoto T (1989) JP Patent 1239114.

Jenkins GM and Kawamura K (1971), Phenolic resin fibres. GB 1228910.

Jiang H, Desai P, Kumar S, Abhiraman AS. Carbon fibers from poly (p-phenylene benzobisthiazole) (pbzt) fibers: conversion and morphological aspects. Carbon 1991;29(4-5):635–44.

Jones HD, Jones TR, Lyle WH. Carbon fiber: results of a survey of process workers and their environment in a factory producing continuous filament. Ann Occup Hyg 1982;26(1–4):861.

Kadla JF, Kubo S. Lignin-based polymer blends: analysis of intermolecular interactions in lignin-synthetic polymer blends. Compos Part A 2005;98:1437–44.

Kadla JF, Kubo S, Gilbert RD, Venditti RA, Compere AL, Griffith WL. Lignin-based carbon fibres for composite fibre applications. Carbon 2002a;40(15):2913–20.

Kadla JF, Kubo S, Gilbert RD, Venditti RA. Hu TQ, editor. Lignin-based carbon fibres, chemical modification, properties, and usage of lignin. New York: Kluwer Academic/Plenum; 2002b. p. 121–37.

Kalback W, Romine E and Bourrat X (1993). Solvated mesophase pitches. US Patent 5259947.

Karacan I, Erdoğan G. An investigation on structure characterization of thermally stabilized polyacrylonitrile precursor fibers pretreated with guanidine carbonate prior to carbonization. Polym Eng Sci 2012;52:937–52.

Kato O, Ito K, Matsunaga K and Sakanishi K (2012) Manufacture of carbon fibres from solid woody materials. JP 2012117162. AIST and Tokai Carbon Co. Ltd.

Kawamura K, Jenkins GM. Mechanical properties of glassy carbon fibres derived from phenolic resin. J Mater Sci 1972;7:1099–112.

Khosravani MR. Composite materials manufacturing processes. Appl Mech Mater 2012;2012(110):1361–7.

Kim UC, Jung JC, Jin SU (2011) Method for manufacturing lignin melt with high processibility. KR 2011108944, 2011, assigned to Kolon Industries.

Kishimoto S and Okazaki S (1977a). Process for producing carbon fibers. US Patent 4009248.

Kishimoto S and Okazaki S (1977b). Process for producing carbon fibers having excellent properties. US Patent 4024227.

Kleinert M, Barth T. Production of biofuel and phenols from lignin by hydrous pyrolysis, In: 15th European biomass conference, Berlin, proceedings 2007; 2007. p. 1297–301.

Knudsen JP. The influence of coagulation variables on the structure and physical properties of an acrylic fiber. Text Res J 1963;33:13–20.

Ko TH, Yieting H, Lin CH. Thermal stabilization of polyacrylonitrile fibers. J Appl Polym Sci 1988;35:631–40. https://doi.org/10.1002/app.1988.070350306.

Kofui Y, Ishida S, Watanabe F, Yoon SH, Wang YG, Mochida I, Kato I, Nakamura T, Sakai Y, Komatsu M. Preparation of carbon fiber from isotropic pitch containing mesophase spheres. Carbon 1997;35:1733–7. https://doi.org/10.1016/S0008-6223(97)00128-0.

Kohn EM (1976). Use of hot buoyant liquid to convert pitch to continuous carbon filament. US Patent 3972968.

Korai Y, Nakamura M, Mochida I, Sakai Y, Fujiyama S. Mesophase pitches prepared from methylnaphthalene by the aid of HF/BF$_3$. Carbon 1991;29:561–7. https://doi.org/10.1016/0008-6223(91)90121-X.

Koslow EE (2007) US Patent 7296691.

Koyama T. Formation of carbon fibers from benzene. Carbon 1972;10:757.

Krutchen CM (1974). Melt extrudable polyacetylene copolymer blends. US Patent 3852235.

Kubo, S. and Kadla, J.F. (2004) Poly(Ethylene oxide)/Organosolv Lignin Blends: Relationship between Thermal Properties, Chemical Structure, and Blend Behavior. Macromolecules, 37, 6904–6911. https://doi.org/10.1021/ma0490552.

Kubo S, Kadla JF. Lignin-based carbon fibres: effect of synthetic polbiocomposite materials, natural fibres, biopolymers, and biocomposites. Compos Part A 2005a;671–97.

Kubo S, Kadla JF. Kraft/lignin/poly(ethylene oxide) blends: effect of lignin structure on miscibility and hydrogen bonding. J Appl Polym Sci 2005b;98:1437–44.

Kubo S, Kadla JF. Lignin-based carbon fibres: effect of synthetic polymer blending on fibre134 properties. J Polym Environ 2005c;13:97–105.

Kubo S, Uraki Y, Sano Y. Thermomechanical analysis of isolated lignins. Holzforschung 1996;50:144–50.

Kubo S, Ishikawa M, Uraki Y, Sano Y. Preparation of lignin fibres from softwood acetic acid lignin relationship between fusibility and the chemical structure of lignin. Mokuzai Gakkaishi 1997;43:655–62.

Kubo S, Uraki Y, Sano Y. Preparation of carbon fibers from softwood lignin by atmospheric acetic acid pulping. Carbon 1998;36:1119–24. https://doi.org/10.1016/S0008-6223(98)00086-4.

Kumar S, Anderson DP, Crasto AS. Carbon fibre compressive strength and its dependence on structure and morphology. J Mater Sci 1993;28:423.

Kuwahara H, Suzuki H and Matsumura S (2008). Polymer for carbon fibre precursor. US Patent 7338997.

Lai C, Zhou Z, Zhang L, Wang X, Zhou Q, Zhao Y, Wang Y, Kadla JF, Kubo S, Gilbert RD, Venditti RA, Compere AL, Griffith WL. Lignin-based carbon fibres for composite fibre applications. Carbon 2002;40(15):2913–20.

Lallave M, Bedia J, Ruiz-Rosas R, Rodriguez-Mirasol J, Cordero T, Otero JC. Filled and hollow carbon nanofibres by coaxial electrospinning of Alcell lignin without binder polymers. Adv Mater 2007;2007(19):4292–6.

Lehmann A, De Ebeling H and Fink HP (2012a) Method for the production of lignin-containing precursor fibres and also carbon fibres WO 2012156443 A1.

Lehmann A, Ebeling H and Fink HP (2012b) Method for the production of lignin-containing precursor fibres and also carbon fibres. International Patent application, WO 2012/156441. Fraunhofer Institute.

Lewis IC (1977). Process for producing carbon fibres from mesophase pitch. US Patent 4032430.

Lin J, Kubo S, Yamada T, Koda K, Uraki Y. Thermostabilized carbon fibers from softwood. Bioresources 2012;7:5634–46.

Lin J, Shang JB, Zhao G. Preparation and characterization of liquefied wood based primary fibres. Carbohydr Polym 2013a;91:224.

Lin L, Li Y, Ko FK. Fabrication and properties of lignin based carbon nanofiber. J Fiber Bioeng Inform 2013b;6:335–47.

Liu W, Zhao G. Effect of temperature and time on microstructure and surface functional groups of activated carbon fibres prepared from liquefied woods. Bioresources 2012;7:5552.

Luo J. Lignin based carbon fibre [thesis master of science, chemical engineering]. USA: University of Maine; 2004.

Luo J, Genco J, Cole B, Fort R. Lignin recovered from the near-neutral hemicellulose extraction process as a precursor for carbon fiber. Bioresources 2011;6:4566–93.

Ma X, Zhao G. Preparation of carbon fibres from liquefied wood. Wood Sci Technol 2010;2010(44):3–11.

Ma X, Zhao GJ. Variation in the microstructure of carbon fibre s prepared from liquefied wood during carbonization. Appl Polym Sci 2011;121:3525.

Ma X, Yuan C, Liu X. Mechanical, microstructure and surface characterizations of carbon fibers prepared from cellulose after liquefying and curing. Materials 2013;2013(7):75–84.

Majibur M, Khan R, Gotoh Y, Morikawa H, Miura M, Fujimori Y, Nagura M. Carbon fiber from natural biopolymer: bombyxmori silk fibroin with iodine treatment. Carbon 2007;45:1035–42.

Maradur SP, Kim CH, Kim SY, Kim B, Kim WC, Yang KS. Preparation of carbon fibres from a lignin copolymer with polyacrylonitrile. Synth Met 2012;162:453–9.

Masahiro T, Takeji O, and Takashi F (1984). Preparation of acrylonitrile precursor for carbon fibre. JP 59204914.

Masaki T, Komatsubara T, Tanaka Y, Nakanishi S, Nakagawa M and Kanamori J (1998). Finish for carbon fiber precursors. US Patent 5783305.

Mavinkurve A, Visser S, Pennings AJ. An initial evaluation of poly(vinylacetylene) as a carbon fiber precursor. Carbon 1995;33(6):757–61.

McCabe M.V (1987). Pretreatment of PAN fiber. US Patent 4661336.

McCorsley CC (1981). Process for shaped cellulose article prepared from a solution containing cellulose dissolved in a tertiary amine N-oxide solvent. US Patent 4246221.

McHenry ER (1977). Process for producing mesophase pitch. US Patent 4026788.

Minus ML, Kumar S. The processing, properties, and structure of carbon fibers. JOM 2005;57(2):52–8.

Minus ML, Kumar S. Carbon fibers. Wiley; 2007a.

Minus ML, Kumar S. Carbon fibre. Kirk-Othmer Encycl Chem Technol 2007b;26:729–49.

Minus et al., (2006) M.L. Minus, H.G. Chae, S. Kumar, Single wall carbon nanotube templated oriented crystallization of poly(vinyl alcohol), Polymer 47 (11) (2006) 3705–3710.

Miyajima N, Akatsu T, Ikoma T, Ito O, Rand B, Tanabe Y, Yasuda E. A role of charge-transfer complex with iodine in the modification of coal tar pitch. Carbon 2000;38:1831–8. https://doi.org/10.1016/S0008-6223(00)00022-1.

Mladenov I, Lyubekeva M. Polyacrylonitrile fibers treated by hydrazine hydrate as a basis for the production of carbon fibers. J Polym Sci A 1983;21:1223–6.

Mochida I, Shimizu K, Korai Y, Otsuka H, Sakai Y, Fujiyama S. Preparation of mesophase pitch from aromatic-hydrocarbons by the aid of HF/BF_3. Carbon 1990;28:311–9. https://doi.org/10.1016/0008-6223(90)90005-J.

Mörck R, Yoshida H, Kringstad KP, Hatakeyama H. Fractionation of Kraft lignin by successive extraction with organic solvents. I. Functional groups, 13C NMR-spectra and molecular weight distributions. Holzforschung 1986;40(Suppl):51–60.

Mörck R, Reimann A, Kringstad KP. Fractionation of Kraft lignin by successive extraction with organic solvents. III. Fractionation of kraft lignin from birch. Holzforschung 1988;42:111–6.

Morgon P. Carbon fibres and their composites. Boca Raton, FL, USA: CRC Press; 2005130–62.

Mukundan T, Bhanu VA, Wiles KB, Johnson H, Bortner M, Baird DG, Naskar AK, Ogale AA, Edie DD, McGrath JE. A photocrosslinkable melt processible acrylonitrile terpolymer as carbon fiber precursor. Polymer 2006;47:4163–71.

Murakami M (1989). Process for producing graphite. US Patent 4876077.

Nagasaka A, Ashitaka H, Kusuki Y, Oda D and Yoshinaga T (1978). Process for producing carbon fibre. US Patent 4131644.

Newell JA, Edie DD. Factors limiting the tensile strength of PBO-based carbon fibers. Carbon 1996;34:551–60.

Norberg I (2012) Carbon fibres from Kraft lignin [doctoral thesis] TRITA-CHE report 2012:13, Stockholm, Sweden, Royal Institute of Technology, p 97 (ISBN 9789175012834)

Norberg I, Nordström Y, Drougge R, Gellerstedt G, Sjöholm E. A new method for stabilizing softwood Kraft lignin fibres for carbon fibre production. J Appl Polym Sci 2012;128(6):3824–30.

Norberg I, Nordström Y, Drougge R, Gellerstedt G, Sjöholm E (2013) A new method for stabilizing softwood kraft lignin fibres for carbon fibre production. J Appl Polym Sci 128:3824–3830

Nordström Y. Development of softwood Kraft lignin based carbon fibres [licentiate thesis]. Department of Engineering Sciences and Mathematics Luleå University of Technology; 2012.

Nordstrom Y, Norberg I, Sjoholm E, Drougge R. A new softening agent for melt spinning of softwood Kraft lignin. J Appl Polym Sci 2012; https://doi.org/10.1002/APP.38795.

Nordström Y, Norberg I, Sjöholm E, Drougge R. A new softening agent for melt spinning of softwood kraft lignin. J Appl Polym Sci 2013;129:1274–9.

Norgren M, Edlund H. Lignin: recent advances and emerging applications. Curr Opin Colloid Interface Sci 2014;19:409–16.

Ohsaki T, Imai K and Miyahara N (1990). Process for preparing a carbon fibre of high strength. US Patent 4925604.

Otani S. On the carbon fiber from the molten pyrolysis products. Carbon 1965;3:31–8. https://doi.org/10.1016/0008-6223(65)90024-2.

Otani S (1982). Dormant mesophase pitch. JP 57100186.

Otani S, Yamada K, Koitabashi T, Yokoyama A. On the raw materials of MP carbon fiber. Carbon 1966;4:425–32. https://doi.org/10.1016/0008-6223(66)90055-8.

Otani S (1982). Dormant mesophase pitch. JP 57100186

Owen PE, Glaister JR, Ballantyne B, Clary JJ. Subchronic inhalation toxicology of carbon fibers. J Occup Med 1986;28(5):373–6.

Ozcan S, Vautard F, Naskar AK. Designing the structure of carbon fibers for optimal mechanical properties. In: Naskar AK, Hoffman WP, editors. Polymer precursor-derived carbon. Washington, DC: ACS Symposium Series; 2014. p. 215–32.

Paiva MC, Kotasthane P, Edie DD, Ogale AA. UV stabilization route for melt-processible PAN-based carbon fibers. Carbon 2003;41:1399–409.

Parry A, Windle A. Carbon fibres from cellulosic precursors: a review. J Mater Sci 2012;47(10):4236–50. https://doi.org/10.1007/s10853-011-6081.

Paul R, Dai X, Hausner A, Naska A, Gallego N. Sintech connect. Available from: 1178194.

Paulauskas FL, White TL, Spruiell JE. Structure and properties of carbon fibres produced using microwave-assisted plasma technology, part 2, In: Proceedings of the international SAMPE technical conference, Long Beach, CA, USA, May 2006; 2006.

Peebles LH. Carbon fibers: formation, structure, and properties. 1st ed. Boca Raton, FL, USA: CRC Press; 19943.

Peebles LH. Carbon fibres: structure and formation. New York, NY, USA: CRC Press; 199512–168.

Peng S, Shao H, Hu X. Lyocell fibers as the precursor of carbon fibers. J Appl Polym Sci 2003;90:1941–7.

Peter S, Beneke H, Oeste F, Fexer W, Jaumann W, Meinbreckse M and Kempfert J (1988). A method for the production of a carbon fiber precursor. US Patent 4756818.

Potter WD, Scott G. Initiation of low-temperature degradation of polyacrylonitrile. Nat Phys Sci 1972;236:32.

Prauchner MJ, Pasa VMD, Otani S, Otani C. Biopitch-based general purpose carbon fibers: processing and properties. Carbon 2005;43(3):591–7.

Qin W, Kadla J. Effect of organoclay reinforcement on lignin-based carbon fibres. Ind Eng Chem Res 2011;50:12548–55.

Qin W, Kadla JF. Carbon fibres based on pyrolytic lignin. J Appl Polym Sci 2012;126:E203–12.

Rangarajan P, Bhanu VA, Godshall D, Wilkes GL, JE MG, Baird DG. Dynamic oscillatory shear properties of potentially melt processable high acrylonitrile terpolymers. Polymer 2002;43:2699–709.

Raskovic V, Marinkovic S. Temperature dependence of processes during oxidation of PAN fibres. Carbon 1975;13:535–8.

Raskovic V, Marinkovic S. Processes in sulfur dioxide treatment of PAN fibers. Carbon 1978;16:351–7.

Richard B, Millington LA and Robert C (1967) Process of graphitizing "polynosic" regenerated cellulose fibrous textile and resulting fibrous graphite textile. US Patent 3322489.

Riggs JP (1972a). Thermally stabilized acrylic fibers produced by sulfation and heating in an oxygen-containing atmosphere. US Patent 3650668.

Riggs JP (1972b). Acrylic fiber stabilization catalyzed by Co(II) and Ce(III) cations. US Patent 3656882.

Riggs JP (1972c). Process for the stabilization of acrylic fibers. US Patent 3656883.

Riggs DM (1985). Process of spinning pitch-based carbon fibers. US Patent 4504454.

Romine E, Rodgers J, Southard M, Nanni E. Solvating component and solvent system for mesophase pitch. US Patent 6717021.

Ruiz-Rosas R, Bedia J, Lalleve M, Loscertales IG, Barrero A, Rodríguez-Mirasol J. The production of submicron diameter carbon fibres by the electrospinning of lignin. Carbon 2010;48:696–705.

Saha D, Li Y, Bi Y, Chen J, Keum JK, Hensley DK, Grappe HA, Meyer 3rd HM, Dai S, Paranthaman MP, Naskar AK. Studies on supercapacitor electrode material from activated lignin-derived mesoporous carbon. Langmuir 2014;30:900–10.

Santangelo JG (1970). Graphitization of fibrous polyamide resinous materials. US Patent 3547584.

Sasaki H and Sawaki T (1990). Method of manufacturing of pitch-base carbon fiber. US Patent 4948574.

Sawaki T, Sasaki H, Nakamura T and Sadanobu J (1989). Process for preparation of high-performance grade carbon fibers. US Patent 4840762.

Sazanov YN, Fedorova GN, Kulikova EM, Kostycheva DM, Novoselova AV, Gribanov AV. Cocarbonization of polyacrylonitrile with lignin. Russ J Appl Chem 2007;80:619. https://doi.org/10.1134/S1070427207040209.

Sazanov YN, Kostycheva DM, Fedorova GN, Ugolkov VL, Kulikova EM, Gribanov AV (2008) Composites of lignin and polyacrylonitrile as carbon precursors. Russ J Appl Chem 81(7):1220–1223.

Schmidt JA, Rye CS, Gurnagul N. Lignin inhibits autoxidative degradation of cellulose. Polym Degrad Stab 1995;49:291–7. https://doi.org/10.1016/0141-3910(95)87011-3.

Scholze B, Hanser C, Meier D. Characterization of the water-insoluble fraction from fast pyrolysis liquids (pyrolytic lignin). J Anal Appl Pyrolysis 2001;60:41–54.

Schurz J. Discoloration effects in acrylonitrile polymers. J Polym Sci 1958;28:438–9.

Seo I, Sakaguchi Y and Kashiwadate K (1989). Process for producing carbon fibers and the carbon fibers produced by the process. US Patent 4863708.

Seo I, Oono T and Murakami Y (1991). Catalytic process for producing raw material pitch for carbon materials from naphthalene. US Patent 5066779.

Sevastyanova O, Qin W, Kadla JF. Effect of nanofillers as reinforcement agents for lignin composite fibres. J Appl Polym Sci 2010;117(5):2877–81.

Seydibeyoglu MO. A novel partially biobased PAN-lignin blend as a potential carbon fibre precursor. J Biomed Biotechnol 2012;2012:598324. https://doi.org/10.1155/2012/598324.

Shen Q, Zhang T and Xu Y (2007) Carbon nanofibres carbonized from lignin/polymer fibres and preparation thereof. CN 101078137.

Shen Q, Zhang T, Zhang WX, Chen S, Mezgebe MJ. Lignin-based activated carbon fibres and controllable pore size and properties. Appl Polym Sci 2011;121:989–94.

Shiedlin A, Marom G, Zillkha A. Catalytic initiation of polyacrylonitrile stabilization. Polymer 1985;26:447–51. https://doi.org/10.1016/0032-3861(85)90210-1.

Shindo A, Fujii R and Souma I (1969). Producing method of carbon or carbonaceous material. US Patent 3427120.

Shiromoto K, Adachi Y and Nabae K (1990). Process for producing carbon fiber. US Patent 4944932.

Singer LS (1977). High modulus, high strength carbon fibers produced from mesophase pitch. US Patent 4005183.

Singer LS. The mesophase and high modulus carbon fibers from pitch. Carbon 1978;16:409–15. https://doi.org/10.1016/0008-6223(78)90085-4.

Sjoholm E, Gellerstedt G, Drougge R, Brodin I (2012) Method for producing a lignin fibre. Application WO 2012/112108 A1, 2012, to be assigned to Innventia AB.

Sliva DE, and Selley W (1975). Continuous method for making spinnable polyacetylene solutions convertible to high strength carbon fibre. US Patent 3928516.

Smith FA, Eckle TF, Osterholm RJ, Stichel RM. Manufacture of coal tar and pitches. In: Hoiberg AJ, editor. Bituminous materials. vol. 3. New York, NY, USA: InterScience Publishers; 1966. p. 57.

Soehngen JW and Willians AG (1977). Carbon fibre production. US Patent 4020145.

Souto F, Calado V, Pereira N. Lignin-based carbon fiber: a current overview. Mater Res Express 2018;5:072001.

Standage A, Matkowshi R. Thermal oxidation of polyacrylonitrile. Eur Polym J 1971;7:775–83.

Sudo K and Shimizu K (1987). Production of lignin-based carbon fiber Patent JP 62-110,922.

Sudo K and Shimizu K (1989) JP Patent 0136618.

Sudo K, Shimizu K. A new carbon-fiber from lignin. J Appl Polym Sci 1992;44:127–34. https://doi.org/10.1002/app.1992.070440113.

Sudo K and Shimizu K (1994). Method for manufacturing lignin for carbon fibre spinning. US Patent 5344921.

Sudo K, Okoshi and Shimizu, K (1988) Am Chem Soc Abstracts V195, 107-Cell.

Sung MG, Sassa K, Tagawa T, Miyata T, Ogawa H, Doyama M, Yamada S, Asai S. Application of a high magnetic field in the carbonizatino process to increase the strength of carbon fibers. Carbon 2002;40:2013–20. https://doi.org/10.1016/S0008-6223(02)00059-3.

Takanori M, Shinya K, Eiichi Y and Otani A (2010) Manufacture of long, ultrafine, and unbranched carbon fibres from lignin derivatives and using fibreization aids. JP 2010242248. Teijin Ltd.

Tang MM, Bacon R. Carbonization of cellulose fibers-I. Low temperature pyrolysis. Carbon 1964;2:211–4. https://doi.org/10.1016/0008-6223(64)90035-1.

Thess A, Lee R, Nikolaev P, Dai H, Petit P, Robert J, Xu C, Lee YH, Kim SG, Rinzler AG, Colbert DT, Scuseria GE, Tomanek D, Fischer J, Smalley RE. Crystalline Ropes of Metallic Carbon Nanotubes. Science 1996;273(5274):483–7.

Thompson SA. Toxicology of carbon fibers. Appl Ind Hyg 1989;12:29–33.

Tibbetts, GG and Devour, MG (1986). Regulation of pyrolysis methane concentration in the manufacture of carbon fibers. US Patent 4565684.

Tiwari S, Bijwe J. Surface treatment of carbon fibers—A review. Procedia Technol 2014;14:505. https://doi.org/10.1016/j.protcy.2014.08.064.

Treacy M, Ebbesen TW, Gibson JM. Exceptionally high Young's modulus observed for individual carbon nanotubes. Nature 1996;381(6584):678–80.

Turner WN and Johnson F.C (1973). Method of manufacturing carbon articles. US Patent 3767773.

Unterweger C, Hinterreiter A, Stifter D, Fuerst C. Cellulose-based carbon fibers: increasing tensile strength and carbon yield, In: Carbon 2018—the world conference on carbon, Madrid, Spain, July 01–06, 2018; 2018.

Uraki Y, Kubo S. Fibrous carbons from woody biomass. Mokuzai Gakkaishi 2006;52(6):337–43.

Uraki Y, Kubo S, Nigo N, Sano Y, Sasaya T. Preparation of carbon fibres from organosolv lignin obtained by aqueous acetic acid pulping. Holzforschung 1995;49:343–50.

Uraki Y, Kubo S, Kurakami H, Sano Y. Activated carbon fibres from acetic acid lignin. Holzforschung 1997;51:188–92.

Uraki Y, Nakatani A, Kubo S, Sano Y. Preparation of activated carbon fibers with large specific surface area from softwood acetic acid lignin. J Wood Sci 2001;47:465–9.

Vakili A, Yue ZR, Fei Y, Cochran H, Allen L, Duran M. Pitch based carbon fiber processing and composites, In: Proceedings of the international SAMPE technical conference, Cincinnati, OH, USA, October 2007; 2007.

Warren CD, Paulauskas FL, Baker FS, Eberle CC, Naskar A. Multi-task research program to develop commodity grade, lower multi-task research program to develop commodity grade, lower cost carbon fibre, In: Proceedings of the SAMPE fall technical conference, Memphis, TN, USA, September 2008; 2008.

Watanabe F, Ishida S, Korai Y, Mochida I, Kato I, Sakai Y, Kamatsu M. Pitch-based carbon fiber of high compressive strength prepared from synthetic isotropic pitch containing mesophase spheres. Carbon 1999;37:961–7. https://doi.org/10.1016/S0008-6223(98)00251-6.

Watt W. Kelly A, Rabotnov YN, editors. Handbook of composites—vol. I. Holland: Elsevier Science; 1985. p. 327–87.

White SM, Spruiell JE, Paulauskas FL. Fundamental studies of stabilization of polyacrylonitrile precursor, part 1: effects of thermal and environmental treatments, In: Proceedings of the international SAMPE technical conference, Long Beach, CA, USA, May 2006; 2006.

Wohlmann B, Woelki M, Ebert A, Engelmann G and Fink HP (2010) Lignin derivative, shaped body comprising the derivative, and carbon fibres produced from the shaped body. WO 2010081775. Toho Tenax Europe GmbH and Fraunhofer Institute of Germany.

Wu Q, Pan D. A new cellulose based carbon fiber from a lyocell precursor. Text Res J 2002;72:405–10.

Wu QL, Gong JH, Zhang ZH, Pan D. Changes in the structure and properties of lyocell fibers treated by a catalyst during pyrolysis. New Carbon Materials 2007;22(1):47–52.

Xiaosong H. Fabrication and properties of carbon fibers. Materials 2009;2:2369–403.

Yamada Y, Honda H and Inoue T (1983). Preparation of carbon fiber. JP 58018421.

Yamada Y, Imamura T, Shibata M, Arita S and Honda H (1986). Method for the preparation of pitches for spinning carbon fibers. US Patent 4606808.

Yang GS, Yoon, JH, Nillesh SL, Kim YC (2011) Lignin based complex precursor for carbon fibres and method for preparing lignin-based carbon fibres using it. KR 2011116604, 2011, assigned to Industry Foundation of Chonnam National University

Yokota H, Kobayashi A, Horikawa J and Miyashita A (1975). Process for producing carbon products. GB 1406378.

Yokoyama A, Nakashima N, Shimizu K. A new modification method of exploded lignin for the preparation of a carbon fiber precursor. J Appl Polym Sci 2003;48:1485–91. https://doi.org/10.1002/app.1993.070480817.

Yoneshiga I and Teranishi H (1970). Japan. Patent specification 2774/701970.

Yoshinori N, Takamaro K, Keitarou F, Takamaro K and Keitaro F (1978). Process for producing carbon fibers. GB 1500675.

Yue ZR, Matthew DP, Vakili A. Fabrication of low-cost composites using roving-like mesophase pitch-based carbon fibers, In: Proceedings of the 17th international conference on composites/nanocomposites engineering, Honolulu, HI, USA, July 2009; 2009.

Zeng F, Pan D, Pan N. Choosing the impregnants by thermogravimetric analysis for preparing rayon-based carbon fibers. J Inorg Org Polym Mater 2005;15:261–7.

Zhang H, Guo L, Shao H, Hu X. Nano-carbon black filled lyocell fiber as a precursor for carbon fiber. J Appl Polym Sci 2006;99:65.

Further reading

Baker FS, Gallego NC, Baker DA. Progress report for lightweighting materials, part 7.A. 2008, Available from: https://www1.eere.energy.gov/vehiclesandfuels/pdfs/lm_08/7_low-cost_carbon_fibre.pdf; 2008.

Baker DA, Harper DP, Rials TG. Carbon fibre from extracted commercial softwood lignin, In: Book of abstracts of the fibre society 2012 fall conferenceBoston: The Fibre Society; 2012a. p. 17–8.

Baker DA, Gallego NC, Baker FS. On the characterization and spinning of an organic-purified lignin toward the manufacture of low-cost carbon fiber. J Appl Polym Sci 2012b;124:227.

Boncher EA, Cooper RN, Everett DH. Preparation and structure of Saran-carbon fibres. Carbon 1970;1970(8):597–605.

Chand S. Review carbon fibers for composites. J Mater Sci 2000;3(5):1303–13.

Chatterjee S, Saito T, Chattacharya P. Lignin-derived carbon fibers. In: Faruk O, Sain M, editors. Lignin in polymer composites. Amsterdam: Elsevier; 2016. p. 207–16.

Chen MCW. Commercial viability analysis of lignin based carbon fibre [master dissertation]. Burnaby, Canada: Simon Fraser University; 2014.

Edie DD. The effect of processing on the structure and properties of carbon fibers. Carbon 1998;36:345–62.

Gellerstedt G, Sjöholm E, Brodin I. The wood-based biorefinery: a source of carbon fiber? Open Agric J 2010;3:119–24.

Gump KHand and Stuetz DE (1977). Stabilization of acrylic fibers and films. US Patent 4004053.

Li Q, Xie S, Serem WK, Naik MT, Liu L, Yuan JS. Quality carbon fibers from fractionated lignin. Green Chem 2017;19(7):1628–34. https://doi.org/10.1039/C6GC03555H.

Otani S, Oya A. Status report on pitch-based carbon fiber in Japan. In: Kawata K, Umekawa S, Kobayashi A, editors. Composites '86: recent advances in Japan and the United States. Tokyo, Japan: Japan Society for Composite Materials; 1986. p. 1–10.

Wu X-F. Free-standing and mechanically flexible mats consisting of electrospun carbon nanofibers made from a natural product of alkali lignin as binder-free electrodes for high-performance supercapacitors. J Power Sources 2014;247:134–41.

Zeng SM, Maeda T, Tokumitsu K, Mondori J, Mochida I. Preparation of isotropic pitch precursors for general purpose carbon fibers (GPCF) by air blowing. II. Air blowing of coal tar, hydrogenated coal tar, and petroleum pitches. Carbon 1993;31(3):413–9.

Zhang M. Carbon fibres derived from dry-spinning of modified lignin precursors [Ph.D. thesis]. Chemical Engineering, Clemson University; 2016.

Zhu D, Xu C, Nakura N, Matsuo M. Study of carbon films from PAN/VGCF composites by gelation/crystallization from solution. Carbon 2002;40(3):363–73.

CHAPTER 3

Types of lignins and characteristics

Lignin is abundantly available and is a natural aromatic resource. In woody plants, it constitutes between 15 and 40 wt% of the dry matter. Lignin is essentially a structural material, imparts strength and rigidity to cell walls, and is quite tolerant to biological attack in comparison with cellulose and other structural polysaccharides (Akin and Benner, 1988; Baurhoo et al., 2008; Kirk, 1971). Plants having a high amount of lignin are found to be more tolerant to direct sunlight and frost (Miidla, 1980).

Presently, lignin is obtained from chemical pulping of wood. Many biomass refineries are now coming on stream, so the lignin produced as a by-product from the production of cellulosic fuel ethanol would represent a valuable feedstock for production of carbon fiber. Lignins produced from organosolv pulping are readily melt spinnable. These lignins have higher purity in comparison with lignins obtained in the chemical pulping process.

In the biosphere the lignin present exceeds 300 billion tonnes and increases by about 20 billion tonnes annually. The term lignin is derived from the Latin word lignum, which means wood, and is present mostly in cell walls of the vascular plants, ferns, and club mosses (Piló-Veloso et al., 1993; McCrady 1991; Gregorováa et al., 2006; Matsushita, 2015; Rosas et al., 2014; Souto et al., 2018).

Lignin does not have a well-defined structure and has special properties such as aromatic nature and lower oxygen content in comparison with polysaccharides. These properties make lignin an attractive feedstock for converting into renewable chemical building blocks and valuable chemicals or materials.

The application potential of lignin very much depends on its availability, proper appreciation of the lignin source, isolation method, and the technical needs of the anticipated applications.

Lignin is a three-dimensional organic polymer. It forms important structural materials in the support tissues of vascular plants. It is quite complex and highly polymerized and is especially common in woody plants. The cellulose walls of the wood get impregnated with lignin. This process is termed lignifications. It significantly increases the strength and hardness of the cell and imparts greater rigidity to the tree. This is very important for the woody plants for standing straight and vertical (Rouhi and Washington, 2001). Natural lignin found in several plants has a complex structure containing both aromatic and aliphatic entities (Nordström, 2012). Importance of lignin has been generally recognized since the early 1900s, though information on lignin has evolved over 1000 years (Glasser et al. 2000).

Our understanding of lignin is limited due to its complex structure. In the recent years, through the application of modern methods of chemical analysis, the lignin field has developed dramatically. This has led to the knowledge of the structure of lignin and also to the applications of lignin. Lignin has been described as a random, three-dimensional network polymer composed of variously linked phenylpropane units. Lignin plays a vital role in providing mechanical support to bind plant fibers together. Lignin also decreases the permeation of water through the cell walls of the xylem, thereby playing an intricate role in the transport of water and nutrients. Finally, lignin plays an important function in a plant's natural defense against degradation by impeding penetration of destructive enzymes through the cell wall.

(Bajpai, 2017; Sarkanen and Ludwig 1971; Sjöström, 1993)

Lignin contains three basic phenylpropane units, originating from p-coumaryl, coniferyl, and sinapyl alcohols (Fig. 3.1). The phenylpropane units are linked via radical coupling reactions for forming a complex three-dimensional macromolecule during biological lignification. The main linkages include β-O-4, β-5, β-b, and 5–5 linkages. The lignin content and composition is found to vary according to species. For instance, softwood lignin mainly contains coniferyl alcohol with small amounts of p-coumaryl alcohol, whereas hardwood lignin contains coniferyl and sinapyl alcohols with small amounts of p-coumaryl alcohol. Grass lignin contains p-coumaryl, coniferyl, and sinapyl alcohols (Matsushita, 2015). Fig. 3.2 shows structure of hardwood and softwood lignin (Gargulak and Lebo, 1999; Nimz, 1974).

Lignin is formed in the cell walls of wood or agricultural crops and plants. Lignin and cellulose work together to provide a structural function in plants. It also forms an effective barrier against attack by insects and fungi. Depending on the specific tree species or plant, the season, the climate, and the plant age, lignin has a variety of specific structures and compositions. The composition of lignin very much depends on the both plant species, in terms of p-hydroxyphenyl (H), guaiacyl (G), and syringyl (S) composition, and also the relative abundance of the chemical linkages in the polymer. Mostly the molecular structure of lignin will be always substantially changed by the process used to separate biomass into its components, thereby resulting in liberation of lignin ("technical lignins"). As a result the characteristics of the technical lignins are different (Table 3.1).

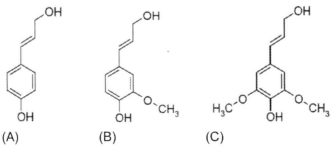

Fig. 3.1 Monolignol monomer species. (A) p-Coumaryl alcohol (4-hydroxyphenyl [H]), (B) coniferyl alcohol (guaiacyl [G]), (C) sinapyl alcohol (syringyl [S]).

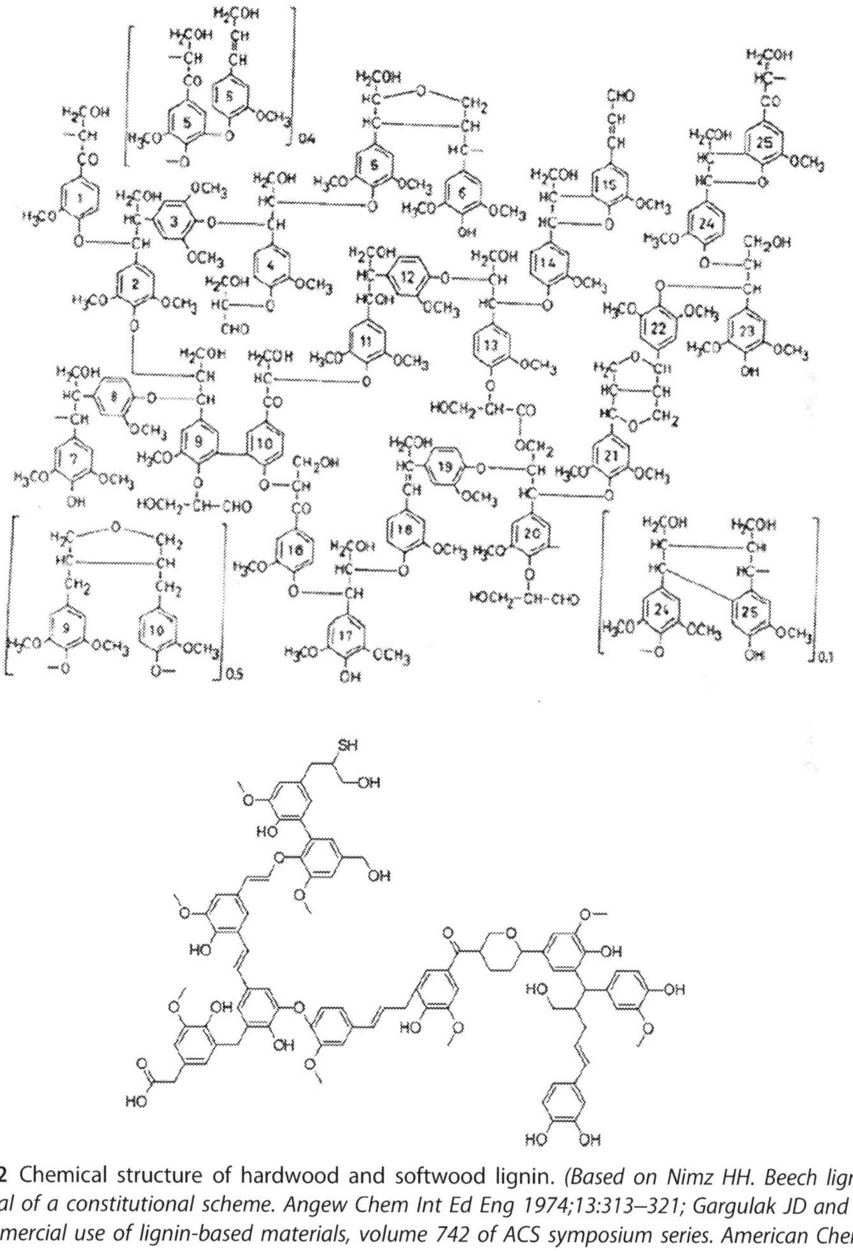

Fig. 3.2 Chemical structure of hardwood and softwood lignin. *(Based on Nimz HH. Beech lignin—proposal of a constitutional scheme. Angew Chem Int Ed Eng 1974;13:313–321; Gargulak JD and Lebo E. Commercial use of lignin-based materials, volume 742 of ACS symposium series. American Chemical Society, 1999. pp. 304–320).*

Table 3.1 The characterization of technical lignins.

Species	Hydroxyl group Total (mmol/g)	Phenolic (mmol/g)	Molecular weight Mw (×10³)	Mn (×10³)	Mw/Mn	References
Kraft lignin						
Softwood	6.5–8.6	2.7–3.5	1.1–45.7	0.5–7.7	2.2–13.4	(Mansson, 1983; El Mansouri and Salvado, 2006, 2007; Ponomarenko et al., 2014; Ekeberg et al., 2006)
Hardwood	6.5–8.4	4.3–4.7	2.4–4.8	0.4–1.3	1.8–12.0	(Mansson, 1983; Ponomarenko et al., 2014; Pan and Saddler, 2013)
Lignosulfonate						
Softwood	Not available	1.2–1.9	10.5–60.2	2.7–6.5	6.7–22.3	(El Mansouri and Salvado, 2007; Ekeberg et al., 2006; Alonso et al., 2001)
Hardwood	Not available	1.4–1.5	6.9–7.8	2.4–4.6	1.7–3.0	(Ekeberg et al., 2006; Alonso et al., 2001; Zhou et al., 2013; Ye et al., 2013)
Ogranosolv lignin						
Softwood (ethanol)	6.3–10	2.7–3.1	2.9–5.4	1.8–3.1	1.6–1.8	(Sannigrahi et al., 2010; Pan et al.,2005)
Hardwood (ethanol)	5.7	2.8	2.0–2.6	1.3–1.6	1.5–1.6	(Pan and Saddler, 2013; Pan et al., 2005)
Miscanthus (formic acid)	3.7–4.9	1.6–2.4	2.8	1.1	2.5	(El Mansouri and Salvado, 2006, 2007; El Mansouri et al., 2012)
Hardwood (acetic acid)	5.5–5.8	3.5–4.0	0.9	Not available	Not available	(Benar et al., 1999; Kin, 1990)
Wheat straw (acetic acid/formic acid)	3.4	1.0	2.2	1.6	1.3	(Delmas et al., 2011)

Table 3.2 Characteristics of the technical lignin.

Molecular weight
Water solubility
Degree of contamination (e.g., remaining covalently bound sugar residues or incorporation of nonnative elements, such as sulfur)
Extent of condensatio
Functional group decoration of the macromolecule

Table 3.3 Some of the major manufacturers of lignins.

Alberta Pacific
Borregaard LignoTech
CIMV
Domtar
Domsjö
Tembec
UPM
Weyerhaeuser

The lignins obtained by various pretreatment methods differ in their chemical structure and also their physicochemical properties.

Currently, lignin is available in huge amounts that can be converted to several other raw materials that have been fractionated, purified, and chemically modified (functionalized). The most important types of lignin are presented in Table 3.2.

Some of the major manufacturers of these products are listed in Table 3.3.

The technical lignins are produced as a by-product of the pulping process. Kraft lignin and lignosulfonate are the major commercial lignins. Other lignins are organosolv lignins. These are obtained from the pulping process using organic solvents like ethanol, acetic acid, and formic acid. The lignin structure differs depending on the pulping process. Also the functional groups and molecular weight vary among different lignins. The functional groups, type of phenylpropane units, linkage between structural units, and molecular weight distributions restrict the applications of lignin as an industrial material.

(Matsushita, 2015)

The technical lignins vary widely in terms of chemical composition (including impurities), molecular structure, molecular weight, and physical properties, such as solubility and hydrophobicity/hydrophilicity. These characteristics will determine to a large extent, which strategies are viable for further valorization.

(Bruijnincx et al., 2016).

3.1 Lignosulfonates

Lignosulfonates are obtained from sulfite pulping processes. This process involves mixing sulfur dioxide with an aqueous solution of base to produce the raw liquor for cooking the wood. In water the sulfur dioxide forms sulfurous acid, which degrades and finally sulfonates the lignin by replacing a hydroxyl group with a sulfonate group, allowing it to be solubilized and separated from the cellulose in nonprecipitated form. The spent sulfite liquor contains lignosulfonate and sugars, mainly monosaccharides that need to be removed or destroyed so as to allow the lignosulfonate to be used effectively as a water-reducing concrete additive.

(Niaounakis, 2015)

The lignosulfonate process is performed in a wide range of pH 2–12 using sulfite with generally either calcium or magnesium as the counterion. The product is soluble in water and in some highly polar organics and amines. Lignosulfonate obtained from sulfite process shows a higher average molecular weight and higher monomer molecular weights as compared with kraft lignin due to incorporation of sulfonate groups on the arenes.

Lignosulfonates are characterized by the sulfonate groups, which are introduced mostly in the α-position of the propyl side chain. This is quite in contrast to sulfonated Kraft lignins, which have the sulfonate on the aromatic ring. As a result the obtained lignin is water-soluble, setting it apart from the other types of technical lignins. The high density of functional groups endows lignosulfonates with unique colloidal properties, accounting for their use as stabilizers, dispersing agents, surfactants, and adhesives. The lignosulfonates have a relatively high molecular weight and high ash content and still contain a significant amount of carbohydrates. Lignosulfonates are commercially available from hardwoods and softwoods.

(www.dutchbiorefinerycluster.nl)

3.2 Kraft lignin

Kraft pulping is an alkaline pulping process. In this process, sodium hydroxide and sodium sulfide are used. It dominates about 96% of the market. Soda and kraft lignins are highly polydisperse having both low-molecular weight phenols and also high–molecular weight lignins attached to carbohydrate residues. Kraft pulping results in substantial degradation of the native lignin structure. It is important to note that kraft lignin has sulfur incorporated in its structure in contrast to soda lignin. The black liquor is burnt to recover chemicals and produce energy and so not available free of cost.

In kraft and soda processes, lignins produced are characterized by more severe fragmentation of the lignin. Earlier efforts to produce carbon fiber from kraft lignin were not successful, particularly, softwood kraft lignin that only produced char on heating (Kubo et al., 1996, 1997). This behavior could be caused by the absence of a main lignin fraction with softening properties, acting as a plasticizer. Highly purified hardwood lignin was used for successfully producing carbon fiber from industrial kraft lignin (Kadla et al., 2002). The lignin was first subjected to heat treatment under vacuum at 145°C for 60 min to enable fiber formation. Volatile components in the lignin were removed, and molecular weight increased.

With the incorporation of small amount of poly(ethylene oxide) as a plasticizer, the spinnability of the lignin improved. Moreover the spinning could be performed at reduced temperature in comparison with that of pure lignin. When poly(ethylene oxide) was added at >10%, self-fusion of the lignin fibers took place. Carbon fiber was produced by thermostabilizing lignin at 250°C for 60 min in air and then carbonizing at 1000°C. The strength properties are presented in Table 3.1. Without using poly (ethylene oxide), but using carefully controlled thermostabilization conditions, including a very slow temperature increase (12°C/h) of the lignin fiber to 250°C, the subsequent carbonization resulted in better strength properties. Further, improvement in strength properties could be achieved by the addition of 5% poly(ethyleneterephtalate) to the lignin before thermostabilization and carbonization (Kubo and Kadla, 2005; www.benthamscience.com).

> The MeadWestVaco Co., the world's largest producer of Kraft lignin, started recovering technical Kraft lignin from the black liquor in 1942, and in 2011 it was the only factory in the world selling commercially technical lignin. Later, to provide a technical Kraft lignin with less impurities, Innventia and Chalmers University Technology developed, together, a process (LignoBoost) that reduces ash content and carbohydrate presence during its recuperation. The technology was sold to Metso Co. FP Innovation, in Canada, also developed a recovering process (Lignoforce) with low ashes, low carbohydrate and low sulfur content. The GreenValue, in Switzerland, also provides nonwood technical soda lignin in commercial scale with low impurity contents.
>
> *(Zhu, 2013; Smolarski, 2012; Gosselink, 2011; iopscience.iop.org)*

LignoBoost technology developed by Innventia and the Chalmers University of Technology in Sweden involves extracting high-purity lignin from a kraft pulp mill. The lignin is precipitated from black liquor by reducing its pH with carbon dioxide. Water is removed from the precipitate using a filter press. Conventional filtration and sodium separation problems are solved by redissolving the lignin in spent wash water and acid. Water is again removed from the resulting slurry and washed with acidified water, for producing pure lignin. All phenols and carboxylic acids get protonated when acidified. Pure lignin with only little contamination of carbohydrates and ash and 2–3 wt% of sulfur is obtained. About half of the sulfur is chemically linked to the lignin.

> Metso acquired the technology in 2008 in its entirety and has developed it further. It sold its first commercial plant for LignoBoost technology at the end of 2011 to Domtar. The unit—at a pulp mill in Plymouth (North Carolina, USA)—started commercial production in 2013 and produces 25.000 tonnes of Kraft lignin. A second plant, which was sold to Stora Enso, produces 50.000 tonnes of dried lignin per year and was started up in the third quarter of 2015 in Sunila, Finland. A similar lignin is commercially produced by Mead-Westvaco in Charleston (USA) with an annual production of about 30,000 tonnes.
>
> *(www.dutchbiorefinerycluster.nl)*

The commercial suppliers of lignin are increasing. But, for high-value application, some purity is needed, and so several technologies are being developed. Lignosulfonates are also available commercially. But they contain impurities, so these are used for low-value applications (Chen, 2014; Gosselink, 2011).

Readily cleavable ether bonds are low in kraft lignin and high in recalcitrant C—C bond linkages (example biphenyl and methylene-bridged ones) and highly condensed. Additionally, sulfur species (mostly as thiols) are incorporated covalently into the structure and therefore constitute important impurities that may affect further valorization (either in material applications or for catalytic depolymerization with sulfur-known poison for several metal catalysts). Kraft lignins have high hydroxyl content and are commercially obtained from hardwoods and softwoods.

3.3 Soda lignin

In soda lignins, no sulfur-containing reagents are used, so these lignins are sulfur free and are different from kraft lignins and lignosulfonates. Because of the relatively severe pulping conditions, lignin becomes condensed and recalcitrant. Like in kraft lignins, soda lignins contain small quantities of ash and carbohydrates, and the purity is low to moderate. Unlike in kraft lignins and lignins obtained under acidic conditions, vinyl ethers are found in soda lignins. Soda lignins are commercially produced from annual crops and also from hardwoods.

3.4 Steam explosion lignin

Steam explosion involves pretreatment of woody biomass with steam at ~200°C or higher and high pressure for short duration followed by fast decompression (Palmqvist et al., 1996; Stenberg et al., 1998; Soederstroem et al., 2004; Krutov et al., 2017). The exploded material is extracted with either aqueous alkali or an organic solvent, and the lignin is obtained as a by-product containing little impurities (carbohydrates and wood extractives). Steam explosion lignin bear a resemblance with the native lignin more than the other produced technical lignins with respect to functional group amount and composition. But the molecular weight of the lignin is substantially reduced.

In the steam explosion process, the biomass is treated with steam to temperatures of 200–220°C. Then, it is rapidly decompressed. The wood is "exploded," and individual fibers and fiber bundles are produced. Extraction with organic solvent or aqueous alkali can be used for isolating lignin in high amount from hardwoods (Robert et al., 1988; Josefsson et al., 2002). By hydrogenolysis of the lignin, subsequent extractions with chloroform and carbon disulfide, and heat treatment and melt spinning, a "green" fiber was obtained. After thermostabilization in air at 210°C, the fibers were subjected to carbonization at 1000°C to produce carbon fiber of "general purpose" grade (Sudo and Shimizu, 1992). The mechanical properties of these fibers are presented in Table 3.1. Similar type of lignin that was subjected to phenolysis followed by purification and heat treatment under vacuum at 280°C was also melt spun, and the resulting fibers were stabilized in air at 300°C and further carbonized at 1000°C. Strength properties were not improved in this case (Table 3.1). However, the yield of the purified lignin was much higher (Sudo et al., 1993).

Further, work using the steam explosion process was conducted by Shimizu et al. (1998). They investigated many hardwood and softwood species and studied the effect of the process parameters for producing lignin suitable for conversion to carbon fiber. "Because of the acidic environment, the degradation and condensation reactions take place in lignin during steam treatment of biomass. It proceeds, through a common carbonium ion intermediate. So, the addition of a low-molecular weight nucleophile, like phenol, will compete with internal nucleophilic centers in the lignin. This results in degradation as the main structural change" (Li et al., 2007).

In this process, molecular weight distribution is similar to the organosolv process. Steam explosion with a sulfur dioxide preimpregnation provides a high fractionation efficiency of alkaline extractable lignin from hardwood, and the efficacy is higher in comparison with the one without preimpregnation (Li et al., 2009). Steam explosion could not give a high efficiency for softwoods with or without the preimpregnation. Two-step steam explosion gives efficiency similar to one-step steam explosion process.

3.5 Organosolv lignins

Organosolv pulping is a fractionation technology which allows for relatively easy isolation of the lignin and comprises a class of processes in which lignocellulosic biomass is treated, often at high temperatures, in a mixture of an organic solvent and water. Ethanol, methanol, acetone, cyclic ethers, or organic acids, such as acetic acid and formic acid or combinations thereof, are often used for organosolv processing. As a result, the cellulose is delignified, and the lignin dissolved in the extraction solvent. The process is acid-catalyzed, with the acid either being produced in-situ from the hemicellulose fraction, or added purposely.

(www.dutchbiorefinerycluster.nl)

The first study using organic solvents for fractionation of wood was conducted in 1893. In this study, ethanol and hydrochloric acid were used (Sidiras and Salapa, 2015). In late 1960s extensive research was conducted on the process (Rinaldi et al., 2016). In 1992 two plants were operating on commercial scale. These were Organocell, ASAM, Acetosolv and Milox (Muurinen, 2000).

Nowadays, many studies are focusing on the use of tetrahydrofuran (THF)-water and γ-valerolactone (GVL)-water cosolvents. The increasing use of these two solvents is because of its ability to partially solubilize/decrystallize crystalline cellulose to sugar platform valorization (Smith et al., 2017). Organosolv activity has grown extensively during the past 50 years, and with the emergence of the biorefineries, its application is expected to increase in the coming years (Rinaldi et al., 2016).

The Alcell process uses aqueous ethanol. It was developed by Repap Enterprises and commercialized in 1989. This process involves delignification using aqueous ethanol. The pH of the cooking liquor is low because of the production of organic acids from hemicellulose. The protolignin can be subjected to condensation reactions between the alpha-position of the side chain and the six-position of another aromatic ring (Pye and Lora, 1991; Sarkanen, 1990; Shimada et al., 1997).

The condensed phenolic groups in solubilized and residual lignin during the Alcell process by 31P NMR were studied. The condensed phenolic hydroxyl content of Alcell lignin (solubilized lignin) was lower compared to kraft lignin as some of the protolignin was condensed during the Alcell process, which led to problems with solubility and pulp retention. Acetic acid has been used for obtaining acetosolv lignins from hardwood, softwood, and grass. The presence of a catalytic amount of a strong acid such as sulfuric acid causes extensive and selective delignification.

(Liu et al. 2000; Kin, 1990; Young and Davis, 1986; Davis et al., 1986; Sano et al., 1989, 1990; Parajó et al., 1993; Shukry et al., 2008; jwoodscience.springeropen.com)

The mechanism of delignification and the structure of acetic acid lignin were determined by the use of model compounds (Davis et al., 1987; Yasuda and Ito, 1987; Yasuda, 1988). "The reactions include acetylation of hydroxyl groups, acidolysis, hydrolysis, and homolytic cleavage of β-aryl ethers, and also elimination of the hydroxyl group at the c-position as formaldehyde. Inter- and intra-molecular condensations also occur during acetic acid pulping. During early 1900s, a delignification process using formic and acetic acid was developed" (Freudenberg, 1959; Erismann et al., 1994).

Biolignin TM was produced by the CIMV (Compagnie Industrielle de la Matière Végétale) process from wheat on a pilot scale (Kin, 1990). Mixture of acetic acid, formic acid, and water was used. This lignin was of low molecular weight, low polydispersity, and contained several free hydroxyl groups.

The structure of the organosolv lignins depends to a large extent on the processing conditions under which they are produced. Organosolv lignins are pure, often with very little carbohydrates and ash impurities. The all-round, general impression that organosolv lignins are also most similar in structure to native lignin and still contains a high fraction of easily cleavable aryl ether linkages (β-O-4) is usually not true, but only holds for organosolv lignins produced under mild conditions in laboratory. While more homogeneous in overall structure, that is, of lower molecular weight and polydispersity, organosolv lignins can be as chemically recalcitrant as soda or Kraft lignins. Organosolv lignins are produced at pilot and demonstration plant scale from hardwoods and softwoods.

(www.dutchbiorefinerycluster.nl; Williamson 1987; Nimz and Casten 1986)

"In the presence of polyethyleneoxide as a plasticizer, Alcell lignin showed an improved spinnability, but the subsequent thermo-stabilization in air had to be done at a very low heating rate for avoiding self-fusing of the filaments. After carbonization, the mechanical properties of the Carbon Fiber improved as compared to other types of Carbon Fiber produced using organosolv lignins" (Table 3.1).

(Kadla et al., 2002)

The spinnability of acetic acid lignin, produced using acetic acid pulping of birch wood, improved after subjecting to thermal treatment at reduced pressure for modifying the lignin structure. An increase in average molecular weight was obtained, whereas the content of methoxyl and acetyl groups remained unchanged. After thermostabilization in air at 250°C and carbonization at 1000°C, carbon fiber with the strength properties presented in Table 3.1 was obtained (Gellerstedt et al., 2010; Uraki et al., 1995). On the other hand a similar approach using softwood did not produce fusible lignin. Carbon fiber could only be obtained after fractionation of lignin for removing the higher-molecular weight materials. Nevertheless, the strength of the resulting carbon fiber was lower in comparison with those produced from hardwoods (Table 3.1) (Kubo et al., 1997, 1998).

Concentrated aqueous solutions of hydrotropic agents (salts that, at high concentration, considerably improve the aqueous solubility of poorly soluble substances) at high temperatures are also used for separating lignin (Gabov et al., 2014). This process was studied from the 1950s through 1980s as an alternative to conventional kraft and sulfite pulping processes (Gabov et al., 2013). The benefits of this process include higher cellulose yield, lower capital costs, heat savings, and simplicity of the process as compared with kraft pulping (Willför and Gustafsson, 2010). Dissimilar to sulfite or kraft pulping that use contaminating inorganic chemicals, the lignin obtained through precipitation by hydrotropic pulping is relatively purer and is found suitable for converting to other chemical products (US Congress, 1989). Studies have shown that hydrotropic lignin resembles with organosolv lignin, which is of high quality (Gabov et al., 2014). But the hydrotropic process is not suitable for softwoods that is the main reason for receiving little attention from the industry. Growing demand for biobased products may bring this option back into consideration, particularly in view of recent efforts in growing woody crops (such as poplar and willow) for energy. Also, attempts are continuing in this area, and research on the use of ionic liquids has been successful for hardwoods and softwoods and also nonwoody biomass (Tan et al., 2009; Cláudio et al., 2015; Muhammad et al., 2012; Mäki-Arvela et al., 2010).

Table 3.4 shows sulfur content and purity of different types of lignins. Fig. 3.3 shows the varieties in physical appearance of a number of typical technical lignins.

Table 3.4 Sulfur content and purity of different types of lignins.

Type of lignin	Scale of production	Separation method	Pretreatment chemistry	Sulfur content	Purity
Kraft	Industrial	Precipitation (pH change) or ultrafiltration	Alkaline	Moderate	Moderate
Soda	Industrial	Precipitation (pH change) or ultrafiltration	Alkaline	Free	Moderate-low
Lignosulfonate	Industrial	Ultrafiltration	Acid	High	Low
Organosolv	Pilot/demo	Dissolved air flotation, precipitation (addition of nonsolvent)	Acid	Free	High
Hydrolysis	Industrial/pilot		Acid	Low-free	Moderate-low
Steam explosion	Demo/pilot		Acid	Low-free	Moderate-low
AFEX	Pilot		Alkaline	Free	Moderate-low

Based on Bruijnincx P, Weckhuysen B, Gruter G and Engelen-Smeets E (2016). Lignin valorisation: the importance of a full value chain approach 22 pages https://www.dutchbiorefinerycluster.nl/.../Lignin_valorisation_-_APC_June_2016.pdf and Mood SH, Golfeshan AH, Tabatabaei M, Jouzani GS, Najafi GH, Gholami M, Ardjm M. Lignocellulosic biomass to bioethanol, a comprehensive review with a focus on pretreatment. Renew Sust Energ Rev 2013;27:77–93.

Fig. 3.3 Photographs showing the differences in physical appearance for a number of typical technical lignins. *(Reproduced with permission (Bruijnincx et al., 2016)).*

References

Akin DE, Benner R. Degradation of polysaccharides and lignin by ruminal bacteria and fungi. Appl Environ Microbiol 1988;54:1117–25.

Alonso MV, Rodríguez JJ, Oliet M, Rodríguez F, García J, Gilarranz MA. Characterization and structural modification of ammonic lignosulfonate by methylolation. J Appl Polym Sci 2001;82:2661–8.

Bajpai P. Carbon fibre from lignin, Springer briefs in material science. Springer (Springer Nature) Switzerland AG; 2017.

Baurhoo B, Ruiz-Feria CA, Zhao X. Purified lignin: nutritional and health impacts on farm animals—a review. Anim Feed Sci Technol 2008;144:175–84.

Benar P, Gonccalves AR, Mandelli D, Schuchardt U. Eucalyptus organosolv lignins: study of the hydroxymethylation and use in resol. Bioresour Technol 1999;68:11–6.

Bruijnincx P, Weckhuysen B, Gruter G and Engelen-Smeets E (2016). Lignin valorisation: the importance of a full value chain approach 22 pages https://www.dutchbiorefinerycluster.nl/…/Lignin_valorisation_-_APC_June_2016.pdf.

Chen MCW. Commercial viability analysis of lignin based carbon fibre (master dissertation). Burnaby, Canada: Simon Fraser University; 2014.

Cláudio A, Neves M, Shimizu K, Lopes J, Freire M, Coutinho J. The magic of aqueous solutions of ionic liquids: ionic liquids as a powerful class of catanionic hydrotropes. Green Chem 2015;17:3948–63.

Davis JL, Young RA, Deodhar SS. Organic acid pulping of wood III. Acetic acid pulping of spruce. Mokuzai Gakkaishi 1986;32:905–14.

Davis JL, Nakatsubo F, Murakami K, Umezawa T. Organic acid pulping of wood IV: reactions of arylglycerol-β-guaiacyl ethers. Mokuzai Gakkaishi 1987;33:478–86.

Delmas GH, Benjelloun-Mlayah B, Bigot YL, Delmas M. Functionality of wheat straw lignin extracted in organic acid media. J Appl Polym Sci 2011;121:491–501.

Ekeberg D, Gretland KS, Gustafsson J, Braten SM, Fredheim GE. Characterisation of lignosulphonates and Kraft lignin by hydrophobic interaction chromatography. Anal Chim Acta 2006;565:121–8.

El Mansouri NE, Salvadó J. Structural characterization of technical lignins for the production of adhesives: application to lignosulfonate, Kraft, soda-anthraquinone, organosolv and ethanol process lignin. Ind Crop Prod 2006;24:8–16.

El Mansouri NE, Salvadó J. Analytical methods for determining functional groups in various technical lignins. Ind Crop Prod 2007;26:116–24.

El Mansouri NE, Vilaseca JF, Salvadó J. Structural changes in organosolv lignin during its reaction in an alkaline medium. J Appl Polym Sci 2012;126:E213–20.

Erismann NM, Freer J, Baeza J, Durán N. Organosolv pulping VII: delignification selectivity of formic acid pulping of Eucalyptus grandis. Bioresour Technol 1994;47:247–56.

Freudenberg K. Biosynthesis and constitution of lignin. Nature 1959;183:1152–5.

Gabov K, Fardim P, da Silva Júnior FG. "Hydrotropic Fractionation of Birch Wood into Cellulose and Lignin: A New Step Towards Green Biorefinery" BioResources 2013;8(3):3518–31.

Gabov K, Gosselink RJ, Smeds AI, Fardim P. Characterization of lignin extracted from birch wood by a modified hydrotropic process. J Agric Food Chem 2014;62(44):10759–67.

Gargulak JD, Lebo E. Commercial use of lignin-based materials, volume 742 of ACS symposium series. American Chemical Society; 1999 p. 304–20.

Gellerstedt G, Sjöholm E, Brodin I. The wood-based biorefinery: a source of carbon fi ber? Open Agric J 2010;3:119–24.

Glasser WG, Northey RA, Schultz TP. Lignin: historical, biological, and materials perspective. Washington, DC: American Chemical Society; 2000.

Gosselink RJA. Lignin as a renewable aromatic resource for the chemical industry [doctoral thesis]. Wegeningen, Netherlands: Wegeningen University; 2011.

Gregorováa A, Košíkováa B, Moravčíkb R. Stabilization effect of lignin in natural rubber. Polym Degrad Stab 2006;31:229–33.

Josefsson T, Lennholm H, Gellerstedt G. Steam explosion of aspen wood. Characterisation of reaction products. Holzforschung 2002;56:289–97.

Kadla JF, Kubo S, Venditti RA, Gilbert RD, Compere A, Griffith W. Lignin-based carbon fibers for composite fiber applications. Carbon 2002;40:2913–20.

Kin Z. The acetolysis of beech wood. TAPPI J 1990;73:237–8.

Kirk TK. Effects of microorganisms on lignin. Annu Rev Phytopathol 1971;9:185–210.

Krutov S, Ipatova E, Vasilyev A. Steam explosion treatments of technical hydrolysis lignin. Holzforschung 2017;71(7–8):571–4.

Kubo S, Kadla JF. Lignin-based carbon fibers: effect of synthetic polymer blending on fiber properties. J Polym Environ 2005;13:97–105.

Kubo S, Ishikawa N, Uraki Y, Sano Y. Preparation of lignin fibers from softwood acetic acid lignin. Relationship between fusibility and the thermal structure of lignin. Mokuzai Gakkaishi 1977;43:655–62.

Kubo S, Uraki Y, Sano Y. Thermomechanical analysis of isolated lignins. Holzforschung 1996;50:144–50.

Kubo S, Uraki Y, Sano Y. Preparation of carbon fibers from softwood lignin by atmospheric acetic acid pulping. Carbon 1998;36:1119–24.

Li J, Henriksson G, Gellerstedt G. Lignin depolymerization and its critical role for delignification of aspen wood by steam explosion. Bioresour Technol 2007;98:3061–8.

Li J, Gellerstedt G, Toven K. Steam explosion lignins; their extraction, structure and potential as feedstock for biodiesel and chemicals. Bioresour Technol 2009;100:2556–61.

Liu Y, Carriero S, Pye K, Argyropoulos DS. A comparison of the structural changes occurring in lignin during Alcell and Kraft pulping of hardwoods and softwoods. In: Grasser WG, Norhey RA, Schultz TP, editors. ACS symposium series 742 lignin: historical, biological, and materials perspectives. Washington, DC: American Chemical Society; 2000. p. 447–64.

Mäki-Arvela P, Anugwom I, Virtanen P, Sjöholm R, Mikkola JP. Dissolution of lignocellulosic materials and its constituents using ionic liquids: a review. Ind Crop Prod 2010;32(3):175–201.

Mansson P. Quantitative determination of phenolic and total hydroxy groups in lignins. Holzforschung 1983;37:143–6.

Matsushita Y. Conversion of technical lignins to functional materials with retained polymeric properties. Wood Sci 2015;61:230–50. https://doi.org/10.1007/s10086-015-1470-2.

McCrady E. The nature of lignin. Alkaline Paper Advocate 1991;4:33–4.

Miidla H. Lignification in plants and methods for its study. In: Regul Rosta Pitan Rast; 1980. p. 87.

Muurinen E. Organsolv pulping—a review and distillation study related to peroxyacid pulping [master dissertation]. Oulu, Finland: University of Oulu; 2000.

Muhammad N, Man Z, Bustam Khalil MA. Ionic liquid—a future solvent for the enhanced uses of wood biomass. Eur J Wood Prod 2012;70:125–33. https://doi.org/10.1007/s00107-011-0526-2.

Niaounakis M. Biopolymers: applications and trends. 1st ed. Elsevier; 2015.

Nimz HH. Beech lignin—proposal of a constitutional scheme. Angew Chem Int Ed Eng 1974;13:313–21.

Nimz HH, Casten R. Chemical processing of lignocellulosics. Holz Roh Werkst 1986;44:207–12.

Nordström Y. Development of softwood Kraft lignin based carbon fibres [licentiate thesis]. Division of Material Science, Department of Engineering Sciences and Mathematics, Luleå University of Technology; 2012.

Palmqvist E, Hahn-Hagerdal H, Galbe M, Larsson M, Stenberg K, Szengyel Z, Tengborg C, Zacchi G. Design and operation of a bench-scale process development unit for the production of ethanol from lignocellulosics. Bioresour Technol 1996;58(2):171–9.

Pan X, Saddler JN. Effect of replacing polyol by organosolv and Kraft lignin on the property and structure of rigid polyurethane foam. Biotechnol Biofuel 2013;6:12–21.

Pan X, Arato C, Gilkes N, Gregg D, Mabee W, Pye K, Xiao Z, Zhang X, Saddler J. Biorefining of softwoods using ethanol organosolv pulping: preliminary evaluation of process streams for manufacture of fuel-grade ethanol and co-products. Biotechnol Bioeng 2005;90:473–81.

Parajó JC, Alonso JL, Vázquez D. On the behavior of lignin and hemicelluloses during the acetosolv processing of wood. Bioresour Technol 1993;46:233–40.

Piló-Veloso D, Nascimento EA, Morais SAL. Isolamento e análise estrutural de ligninas. Quim Nova 1993;16:435–48.

Ponomarenko J, Dizhbite T, Lauberts M, Viksna A, Dobele G, Bikovens O, Telysheva G. Characterization of softwood and hardwood LignoBoost Kraft lignins with emphasis on their antioxidant activity. Bioresources 2014;9:2051–68.

Pye EK, Lora JH. The AlcellTM process. A proven alternative to Kraft pulping. TAPPI J 1991;74:113–8.

Rinaldi R, Jastrzebski R, Clough MT, Raplh J, Kennema M, Bruijnincx PCA, Weckhuysen BM. Paving the way for lignin valorisation: recent advances in bioengineering, biorefining, catalysis. Angew Chem Int Ed 2016;55(29):8164–215.

Robert D, Bardet M, Lapierre C, Gellerstedt G. Structural changes in aspen lignin during steam explosion treatment. Cell Chem Technol 1988;22:221–30.

Rosas JM, Berenguer R, Valero-Romero MJ, Rodríguez-Misarol J, Cordero T. Preparation of different carbon materials by thermochemical conversion of lignin. Front Mater 2014;1:1–17.

Rouhi AM, Washington C. Only facts will end lignin war. Sci Technol 2001;79(14):52–6.

Sannigrahi P, Ragauskas AJ, Miller SJ. Lignin structural modifications resulting from ethanol organosolv treatment of loblolly pine. Energy Fuel 2010;24:683–9.

Sano Y, Maeda H, Sakashita Y. Pulping of wood at atmospheric pressure I: pulping of hardwoods with aqueous acetic acid containing a small amount of organic sulfonic acid. Mokuzai Gakkaishi 1989;35:991–5.

Sano Y, Nakamura M, Shimamoto S. Pulping of wood at atmospheric pressure II: pulping of birch of wood with aqueous acetic acid containing a small amount of sulfuric acid. Mokuzai Gakkaishi 1990;36:207–11.

Sarkanen KV. Chemistry of solvent pulping. TAPPI J 1990;73(10):215–9.

Sarkanen KV, Ludwig CH. In: Sarkanen KV, Ludwig CH, editors. Lignin: occurrence, formation, structure and reactions. New York: Wiley-Interscience; 1971. p. 916.

Shimada K, Hosoya S, Ikeda T. Condensation reactions of softwood and hardwood lignin model compounds under organic acid cooking conditions. J Wood Chem Technol 1997;17:57–72.

Shimizu K, Sudo K, Ono H, Ishihara M, Fujii T, Hishiyama S. Integrated process for total utilization of wood components by steam explosion pre-treatment. Biomass Bioenergy 1998;14:195–203.

Shukry N, Fadel SM, Agblevor FA, El-Kalyoubi SF. Some physical properties of acetosolv lignins from bagasse. J Appl Polym Sci 2008;109:434–44.

Sidiras DK, Salapa IS. Engineering conferences international. Available from: http://dc.engconfintl.org/cgi/viewcontent.cgi?article=1013&context=biorefinery_I; 2015. (Accessed 20 February 2018).

Sjöström E. Wood chemistry: fundamentals and application. Orlando: Academic Press; 1993 p. 293.

Smith MD, Cheng X, Petridis LP, Mostofian B, Smith JC. Organosolv-water cosolvent phase separation on cellulose and its influence on the physical deconstruction of cellulose: a molecular dynamics analysis. Sci Rep 2017;7:14494.

Smolarski N. High-value opportunities for lignin: unlocking its potential. Available from: http://greenmaterials.fr/wpcontent/uploads/2013/01/High-value-Opportunities-for-Lignin-Unlocking-its-Potential-Market-Insights.pdf; 2012.

Soederstroem J, Galbe M, Zacchi G. Effect of washing on yield in one- and two-step steam pretreatment of softwood for production of ethanol. Biotechnol Prog 2004;20(3):744–9.

Souto F, Calado V, Pereira N. Lignin-based carbon fiber: a current overview. Mater Res Express 2018;5:072001.

Stenberg K, Tengborg C, Galbe M, Zacchi G. Optimization of steam pretreatment of SO_2-impregnated mixed softwoods for ethanol production. J Chem Technol Biotechnol 1998;71(4):299–308.

Sudo K, Shimizu K. A new carbon fiber from lignin. J Appl Polym Sci 1992;44:127–34.

Sudo K, Shimizu K, Nakashima N, Yokoyama A. A new modification method of exploded lignin for the preparation of a carbon fiber precursor. J Appl Polym Sci 1993;48:1485–91.

Tan SSY, MacFarlane DR, Upfal J, Edye LA, Doherty WOS, Patti AF, Pringle JM, Scott JL. Extraction of lignin from lignocellulose at atmospheric pressure using alkylbenzenesulfonate ionic liquid. Green Chem 2009;11:339–45.

U.S. Congress, Office of Technology Assessment. Technologies for reducing dioxin in the manufacture of bleached wood pulp. OTA-BP-O-54. Washington, DC: U.S. Government Printing Office; 1989.

Uraki Y, Kubo S, Nigo N, Sano Y, Sasaya T. Preparation of carbon fibers from organosolv lignin obtained by aqueous acetic acid pulping. Holzforschung 1995;49:343–50.

Williamson PN. Repap's ALCELL process: how it works and what it offers. Pulp Pap Can 1987;88(12):47.

Willför S, Gustafsson J. "The Forest Based Biorefinery: Chemical and Engineering Challenges and Opportunities", Accessed December 2015. http://web.abo.fi/instut/pcc/presentations_pdf/Willf%C3%B6r_Gustafsson_Lignin.pdf; 2010.

Yasuda S. Behavior of lignin in organic acid pulping II: reaction of phenylcoumaran and 1,2-diaryl-1,3-propanediol with acetic acid. J Wood Chem Technol 1988;8:155–64.

Yasuda S, Ito N. Behavior of lignin in organic acid pulping I: reaction of arylglycerol-b-aryl ethers with acetic acid. Mokuzai Gakkaishi 1987;33:708–15.

Ye DZ, Zhang MH, Gan LL, Li QL, Zhang X. The influence of hydrogen peroxide initiator concentration on the structure of eucalyptus lignosulfonate. Int J Biol Macromol 2013;60:77–82.

Young RA, Davis JL. Organic acid pulping of wood. Part II acetic acid pulping of aspen. Holzforschung 1986;40:99–108.

Zhou H, Yang D, Qiu X, Wu X, Li Y. A novel and efficient polymerization of lignosulfonates by horseradish peroxidase/H_2O_2 incubation. Appl Microbiol Biotechnol 2013;97:10309–20.

Zhu W. Equilibrium of lignin precipitation—the effects of pH, temperature, ion strength and wood origins [Bacharel degree]. Goethenburg, Sweden: Chalmers University of Technology; 2013.

Further reading

Mood SH, Golfeshan AH, Tabatabaei M, Jouzani GS, Najafi GH, Gholami M, Ardjm M. Lignocellulosic biomass to bioethanol, a comprehensive review with a focus on pretreatment. Renew Sust Energ Rev 2013;27:77–93.

CHAPTER 4

Conversion of lignin to carbon fiber

Various spinning methods are used for fiber formation from polymers (Dallmeyer et al., 2010; Thunga et al., 2014; Zhang and Ogale, 2014). These are presented in Table 4.1.

4.1 Melt spinning

Wet spinning is the oldest process. The process is called wet spinning as the solution is squeezed out directly into the precipitating liquid. Wet spinning of fiber is a kind of solution spinning where polymer is dissolved in a suitable solvent and is squeezed out through spinneret into a solvent-nonsolvent mixture (coagulant). Because of mutual diffusion of solvent and nonsolvent, polymer solution coagulates and the fibers are produced. The coagulated fiber, is subjected to washing, which is done in several stages for removing the trapped solvent. It is then stretched under wet and dry condition for achieving desired fiber denier. The fiber is also coated with a spin finish solution for improving its hand lability (www.nal.res.in).

4.2 Dry spinning

This method involves dissolving the polymer in to a suitable solvent for making the fiber solution. "This solution is then extruded under heat and pressure into an air gap before it enters a coagulation bath. The produced fibre is then washed and dried before it is heat treated and drawn. This is an alternative method to wet spinning and is required as spinning directing into the bath, for some fibres, creates microvoids that negatively affect the fibre properties, this is due to the solvent being drawn out of the liquid too quickly. An inert atmosphere may be required to prevent oxidisation in some polymers, if so fibres are extruded into a nitrogen atmosphere" (www.tikp.co.uk).

This method is generally needed for high-performance fibers with a liquid crystal structure. Because of their structural properties, their melt temperature is identical as or closer to their decomposition temperature, so they should be dissolved in proper solvent and extruded in this way (www.tikp.co.uk).

Table 4.1 Most common spinning techniques.

Melt spinning
Dry spinning
Wet spinning
Dry-jet wet spinning
Electrospinning

4.3 Wet spinning

In dry spinning, solidification is obtained by evaporating the solvent in inert gas or air stream instead of precipitating the polymer by chemicals or by dilution. The filaments do not get contacted with a precipitating liquid. So, there is no requirement for drying, and the solvent recovery becomes easy.

> *Dry spinning is used to form polymeric fibers from solution. The polymer is dissolved in a volatile solvent and the solution is pumped through a spinneret (die) with numerous holes (one to thousands). As the fibers exit the spinneret, air is used to evaporate the solvent so that the fibers solidify and can be collected on a take-up wheel. Stretching of the fibers provides for orientation of the polymer chains along the fiber axis. Cellulose acetate (acetone solvent) is an example of a polymer which is dry spun commercially in large volumes. Due to safety and environmental concerns associated with solvent handling, this technique is used only for polymers which cannot be melt spun.*
>
> **(www.polymerprocessing.com)**

4.4 Dry-jet wet spinning

Melt spinning method is fast. It is commonly used for forming polymeric fibers. The fiber-forming substance is melted and extruded using the spinneret. Then, it is directly solidified by cooling. With this method the use of solvents can be substantially reduced. The polymer is melted to produce a liquid spinning solution or dope. The polymers are fed to a hopper, which is heated. There is a grid at the bottom that allows the molten liquid to pass through. Then the solution is purified by filter. The molten polymer is extruded at a constant rate and high pressure through a spinneret into a comparatively cooler air stream. The filaments get solidified. Finally the filament yarn is instantly wound onto bobbins or further treated for certain desired properties or end application.

Otani in 1969 reported melt spinning of lignin. He described many techniques of forming fiber from lignin using a one-pot melt spinning technique. Since that time, several achievements arose, and recent technical developments allow faster and easier handling (Otani et al., 1969).

Baker et al. (2012) were able to melt spun lignin fiber from hardwood kraft lignin and also an organic purified hardwood lignin. An organic solvent purification process was used for removing the impurities from lignin for improving its melt spinnability. The fibers were then stabilized and carbonized for obtaining lignin-based carbon fiber. The major disadvantage was the slow heating rates for fiber stabilization.

Oxidative stabilization can only be obtained at heating rates smaller than 0.05°C/min. It was found that reducing the rate of heating significantly increased the glass transition temperature and also enabled the lignin to cross-link and produce stabilized fibers. The strength properties of the carbon fibers were inferior, and so, research is needed to reduce stabilization time and improve the properties.

4.5 Electrospinning

Electrospinning is an alternative to the melt spinning method. Electric force is used in this method for drawing charged threads of polymer solutions, or polymer melts up to produce ultrafine fibers. This method shares characteristics of both electrospraying and traditional dry spinning of fibers. This method does not need high temperatures or the use of coagulants for producing solid threads from solution. This process is especially suitable for the production of fibers using complex and large molecules. Electrospinning from molten precursors is also practiced; this method makes certain that no solvent is carried over into the final product.

> *All these methods involve pumping the melt or solution of the polymer through holes in a spinneret. The spinneret holes match the desired filament count of the carbon. Melt spinning is not possible in case of Polyacrylonitrile because Polyacrylonitrile decomposes below its melting temperature. Polyacrylonitrile based fibres are mostly produced by either wet spinning or melt assisted spinning process. In both the processes organic solvent is used. In case of melt assisted spinning, a solvent in the form of hydrating agent to decrease the melting point of Polyacrylonitrile is used. The polymer can be melted with a low melting point and pumped through a spinneret having several holes. In case of wet spinning, the dissolved precursor is immersed in a liquid coagulation bath and extruded through holes in a spinneret. The wet-spun fibre is drawn by using rollers through a wash for removing the excess coagulant. The fibres are then dried and stretched to the correct fibre specification.*
>
> **(McConnell, 2008; Chung, 1994).**

The direct melt spinning is most commonly used for the lignin fiber (Souto et al., 2018). The process needs the extrusion of pure polymer precursor directly into fiber form, eliminating the need of extra expenses of recovering solvent and providing a more environment friendly option. Lignin must have higher purity (more than 99%) for melt spinning. The particle size should not be larger than 1 μm (Eberle, 2012). Lignin should be produced in such a way that allows low melt flow temperature so that it can be melt spun without getting polymerized during extrusion. High glass transition temperature is also needed for the fiber to get stabilized at an acceptable rate. The glass transition temperature is maintained above the oxidation temperature in order that the fiber cross-links and gets stabilized without infusion. Following conditions should be precisely controlled for producing carbon fiber having desired strength properties:
- spinning conditions
- treatment temperatures
- temperature ramping profiles

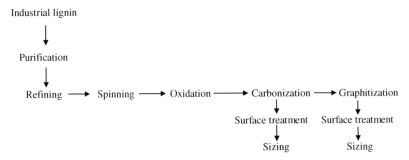

Fig. 4.1 Production steps involved in the production of carbon fiber from lignin. *(Based on Baker DA, Rials TG. Recent advances in low-cost carbon fibre manufacture from lignin. J Appl Polym Sci 2013;130:713–28; Chen MCW. Commercial viability analysis of lignin based carbon fibre [Master thesis]. Simon Fraser University; 2014.)*

Production steps involved in carbon fiber production from lignin are shown in Fig. 4.1.

Electrospinning [sic] produces fibres in the diameter from 0.03 mm to 1 mm range and is an effective method for the fabrication of fibres in submicron diameter range, 0.1 μm, from polymer solutions or melts. Generally, an electric potential is used between a droplet of polymer solution at the end of a capillary and a grounded collector. When the applied electric field overcomes the surface tension of the droplet, a charged jet of polymer solution or melt is produced. The jet grows longer and thinner until it gets solidified or collects on the collector.

(Bajpai, 2017)

Following properties control the fiber morphology (Lin et al., 2013):
- solution conductivity
- concentration
- viscosity
- molecular weight
- applied voltage

Lignin is a thermoplastic polymer making melt spinning is an applicable method to spin fibres. In contrast to kraft lignin, organosolv lignin contains only a very small amount of inorganic material, which provides good melt spinning opportunities. Inorganic materials are contaminants that will degrade carbon fibre properties and are a result of the lignin recovery process.

(Bajpai, 2017)

Dry-spinning was done over a range of concentrations (1.85 to 2.15 g/mL acetone) temperatures range of 25–50°C. All of the resulting dry-spun lignin fibres showed a crenulated surface pattern, with increased crenulation obtained for fibres spun at higher temperatures. Presence of some doubly-convex and sharp crevices was observed on fibres produced from solutions containing reduced concentrations (1.85 and 2.00 g lignin/mL solvent). Contrary to this, no crevices were seen on the fibres obtained from the concentrated solution (2.15 g/mL), possibly because of the reduced extent of solvent out-diffusion. Dry-spinning at room temperature was conducted to obtain fibres with relatively smooth surface, but the pressure drop was very high. These results clearly establish temperature/concentration combinations for dry-spinning of Ace-SKL. About 30% higher surface area was obtained in the crenulated lignin fibres (as compared with equivalent circular fibres), showing the potential benefits of such biomass-derived fibres in providing larger fibre/matrix bonding area when used in composites.

(Zhang, 2016; Bajpai, 2017)

Lignin is melt spun into fiber under an inert atmosphere after it is oxidatively thermo-stabilized and carbonized.

Oxidation (stabilization)

Fiber should be chemically changed from the linear bonding to ladder bonding structure before it reaches the carbonizing stage. This process is termed oxidation, which is conducted by heat. The spool is passed through oxidation ovens where temperature is in the range from 200°C to 300°C. The process combines oxygen molecules from the air with the fiber causing the polymer chains to cross-link and thereby increasing the fiber density. In the case of polyacrylonitrile-based fiber, the density can increase from ~1.18 to ~1.38 g/cc. During the heating process the fiber produces heat; therefore the temperature of the oven and airflow has to be precisely controlled and monitored for avoiding overheating. The oxidation process takes from 30 to 120 min in the case of polyacrylonitrile-based fiber (Cavette, 2006; McConnell, 2008).

Carbonization

After stabilization the fibers are heated at a temperature of ~1000–3000°C for several minutes in a furnace that is filled with an inert gas mixture that do not contain any oxygen. The absence of oxygen prevents the fibers from, burning (Zoltek, n.d.). The gas pressure inside the furnace is kept higher in comparison with the outside air pressure, and the oven is sealed so that oxygen and air do not enter. When the fibers are heated, they start to loose noncarbon atoms and many impurities in the form of gases like water vapor, ammonia, carbon monoxide, carbon dioxide, hydrogen, and nitrogen. As the noncarbon atoms are removed, the remaining carbon atoms get tightly bonded along the axis of the fiber. Sometimes, two separate furnaces are used for carbonizing the fiber at different temperatures for controlling the process. After carbonization the fiber contains more than 90% carbon depending on the process (Cavette, 2006; www.zoltek.com; McConnell, 2008).

Graphitization

The terms carbon and graphite are often used interchangeably by mistake. In reality, they are different. Carbon denotes fibres carbonized at about 1,315°C and that contain 93% to 95% carbon. Graphite is graphitized at 1,900°C to 2,480°C and contains more than 99% elemental carbon. Graphitization [sic] step is not always included after carbonization. Its usage depends on the final usage of the produced carbon fibre.

(McConnell, 2008)

Surface treatment

After carbonization the fibers develop surfaces that do not bond well with the epoxies and other materials. For giving the fibers better bonding properties, the surface should be slightly oxidized and treated for introducing small surface roughness that also improves the mechanical bonding. The oxidation is obtained by submerging the fibers in gases (air, carbon dioxide, or ozone) or in different types of liquids (hypochlorite or nitric acid). Surface treatment of the fibers can also be performed using electrolyte. The surface treatment process is generally precisely controlled for avoiding defects on surface that may cause fiber failure (Cavette, 2006). "Surface treatment is generally done for providing a better interaction between the fiber and polymeric matrix as a guarantee of quality for composites. It oxidizes the fiber surface, creates some roughness and includes functional groups" (http://zoltek.com/carbon-fiber/how-is-carbon-fiber-made/).

Sizing

After the surface treatment, the fibres are coated in order to protect from damage during winding or weaving. Coating materials are selected to be compatible with the adhesive later in the manufacturing process. Typical coating materials include epoxy, polyester, nylon and urethane. The coated fibres are loaded into spinning machine and twisted into tows with various sizes.

(Cavette, 2006; McConnell, 2008).

An electrolytic coating is also a possible oxidative process (Park and Heo, 2015). Several companies kept these procedures confidential (Park and Heo, 2015). Sizing protects the physical properties of the fiber, lubricates it, and makes easier to handle when they are still brittle at this stage (Park and Heo, 2015; http://zoltek.com/carbon-fiber/how-is-carbon-fiber-made/). "Coating must be compatible with the matrix resin. Parameters at this stage [sic] are crucial in the carbon fiber specification" (Park and Heo, 2015; Norberg, 2012).

The integrity of the lignin fibre during this process depends on its ability to cross-link, so that the glass transition of the material is maintained above the process temperature, eventually rendering it infusible. The process is quite complex. Careful control of the following conditions are needed for obtaining the carbon fibre of superior strength:

- *Lignin spinning conditions*
- *Treatment temperatures*
- *Ramping profiles*

Lignin, which is naturally partially oxidized, needs critical control of the melt-spinning step. The lignin should be produced in such a way that it has low enough melt flow temperature for it to be melt spun without getting polymerized during extrusion, but a high enough glass transition for fibre stabilization to proceed at an acceptable rate. Therefore opportunity exists for producing carbon fibre from lignin by melt extrusion.

(Bajpai, 2017)

The differences in structure between hardwood lignin and softwood lignin affect the cross-linking ability of lignin and the spinning properties in the carbon fiber production process. In case of softwood kraft lignin, the difficulties of spinning and thermal mobility have been observed (Kadla et al., 2002; Norberg et al., 2013; Hosseinaei et al., 2016). However, a condensed structure (C-C linkages) and unoccupied C5 sites in lignins with a high guaiacyl unit content (grasses and softwood) result in faster thermostabilization and a reduction in production time of carbon fiber (Norberg et al., 2013; Hosseinaei et al., 2016).

"During the oxidative stabilization, the major reactions in lignin are oxidative which are caused by radicals (introduced by homolytic cleavage of the β-O-4' bond during heat treatment) and rearrangement reactions. During the carbonization step, the stabilized fiber is converted into a Carbon Fiber, through the elimination of all elements except carbon" (Braun et al., 2005). During this process the formation of the graphite structure of the lignin-based carbon fiber was established (Mainka et al., 2015).

"Alternatives have been developed to potentiate the Carbon Fiber production from lignin, mostly related to the improvement of spinning process, example chemical modification through the esterification of lignin phenolic groups, hydrogenation, and phenolation, plasticizing lignin through blending with synthetic polymers, such as polyethylene oxide, polyethylene terephthalate Polyacrylonitrile, and reinforcement with clay or carbon nanotubes" (Sudo and Shimizu, 1992; Sudo et al., 1993; Uraki et al., 2001; Kadla and Kubo, 2004; Kubo and Kadla, 2005; Sevastyanova et al., 2010; Thunga et al., 2014; Ding et al., 2016; Liu et al., 2016; Wang et al., 2016; Youe et al., 2016). Furthermore, lignin having improved melt spinning ability and high thermal mobility is obtained by purification and fractionation.

For improving the homogeneity and reducing the complexity of lignin, fractionation methods by an eluotropic series of organic solvents and precipitation by the pH effect are potential options. Also, the use of membranes has been thoroughly studied. These methods allow lignin to be produced with more uniform molecular weights, thus improving the Carbon Fiber spinning ability and performance.

(Brodin et al., 2009; Baker et al., 2012; Nordström et al., 2013; Wang and Chen, 2013; Li and McDonald, 2014; Duval et al., 2015; Kleinhans and Salmén, 2016)

A new method for fractionating and modifying kraft lignin using an oxidation process for treating kraft lignin with laccase and a mediator was reported (Li et al., 2017). Lignin fractions having different molecular weights, functional groups, and interunitary linkages were obtained. The insoluble fraction of the treated lignin was used with polyacrylonitrile at a ratio of 1:1 as the starting material for production of carbon fiber. Spinnability was improved because of the better miscibility. The elastic modulus of lignin-polyacrylonitrile carbon fiber matched with those of several commercial carbon fibers (polyacrylonitrile-carbon fiber).

References

Bajpai P. Carbon fibre from lignin. In: SpringerBriefs in material science. Springer: (Springer Nature Switzerland AG); 2017.

Baker DA, Gallego NC, Baker FS. On the characterization and spinning of an organic-purified lignin toward the manufacture of low-cost carbon fiber. J Appl Polym Sci 2012;124(1):227–34. https://doi.org/10.1002/app33596.

Braun JL, Holtman KM, Kadla JF. Lignin-based carbon fibers: oxidative thermostabilization of kraft lignin. Carbon 2005;43(2):385–94. https://doi.org/10.1016/j.carbon.204.09.027.

Brodin I, Sjöholm E, Gellerstedt G. Kraft lignin as feedstock for chemical products: the effects of membrane filtration. Holzforschung 2009;63(3):290–7. https://doi.org/10.1515/HF.2009.049.

Cavette C. How products are made: carbon fiber, May 10, 2014. Retrieved from: http://www.madehow.com/Volume-4/Carbon-Fiber.html; 2006.

Chung DDL. Carbon fibre composites. Butterworth-Heinemann: Oxford; 1994.

Dallmeyer I, Ko F, Kadla JF. Electrospinning of technical lignins for the production of fibrous networks. J Wood Chem Technol 2010;30(4):315–29. https://doi.org/10.1080/02773813.2010.527782.

Ding R, Wu H, Thunga M, Bowler N, Kessler MR. Processing and characterization of low-cost electrospun carbon fibers from organosolv lignin/polyacrylonitrile blends. Carbon 2016;100:126–36. https://doi.org/10.1016/j.carbon.2015.12.078.

Duval A, Vilaplana F, Crestini C, Lawoko M. Solvent screening for the fractionation of industrial kraft lignin. Holzforschung 2015;70(1):11–20. https://doi.org/10.1515/hf-2014-0346.

Eberle C. Carbon fibre from lignin, Oak Ridge, TN. Retrieved from: http://www.cfcomposites.org/PDF/Breakout_Cliff.pdf: ORNL; 2012.

Hosseinaei O, Harper DP, Bozell JJ, Rials TG. Role of physicochemical structure of organosolv hardwood and herbaceous lignins on carbon fiber performance. ACS Sustain Chem Eng 2016;4(10):5785–98. https://doi.org/10.1021/acssuschemeng.6b01828.

Kadla JF, Kubo S. Lignin-based polymer blends: analysis of intermolecular interactions in lignin–synthetic polymer blends. Compos A: Appl Sci Manuf 2004;35(3):395–400. https://doi.org/10.1016/j.compositesa.2003.09.019.

Kadla JF, Kubo S, Venditti RA, Gilbert RD, Compere AL, Griffith W. Lignin-based carbon fibers for composite fiber applications. Carbon 2002;40(15):2913–20. https://doi.org/10.1016/S0008-6223(02)00248-8.

Kleinhans H, Salmén L. Development of lignin carbon fibers: evaluation of the carbonization process. J Appl Polym Sci 2016;133(38). https://doi.org/10.1002/app.43965.

Kubo S, Kadla JF. Lignin-based carbon fibers: effect of synthetic polymer blending on fiber properties. J Polym Environ 2005;13(2):97–105. https://doi.org/10.1007/s10924-005-2941-0.

Li H, McDonald AG. Fractionation and characterization of industrial lignins. Ind Crop Prod 2014;62:67–76. https://doi.org/10.1016/j.indcrop.2014.08.013.

Li J, Wang M, She D, Zhao Y. Structural functionalization of industrial softwood kraft lignin for simple dip-coating of urea as highly efficient nitrogen fertilizer. Ind Crop Prod 2017;109:255–65. https://doi.org/10.1016/j.indcrop.2017.08.011.

Lin L, Yingjie L, Ko FK. Fabrication and properties of lignin based carbon nanofibre. J Fibre Bioeng Inform 2013;6(4). https://doi.org/10.3993/jfbi12201301.

Liu HC, Chien A-T, Newcomb BA, Davijani AAB, Kumar S. Stabilization kinetics of gel spun polyacrylonitrile/lignin blend fiber. Carbon 2016;101:382–9. https://doi.org/10.1016/j.carbon.2016.01.096.

Mainka H, Täger O, Körner E, Hilfert L, Busse S, Edelmann FT, Herrmann AS. Lignin—an alternative precursor for sustainable and cost-effective automotive carbon fiber. J Mater Res Technol 2015;4(3):283–96. https://doi.org/10.1016/j.jmrt.2015.03.004.

McConnell V. The making of carbon fibre, CompositesWorld 2008; December 19, May 10, 2014. Retrieved from: http://www.compositesworld.com/articles/the-making-of-carbonfibre.

Norberg I. Carbon fibres from kraft lignin. [Ph.D. thesis]Stockholm, Sweden: KTH Royal Institute of Technology; 2012.

Norberg I, Nordström Y, Drougge R, Gellerstedt G, Sjöholm E. A new method for stabilizing softwood kraft lignin fibers for carbon fiber production. J Appl Polym Sci 2013;128(6):3824–30. https://doi.org/10.1002/app.38588.

Nordström Y, Norberg I, Sjöholm E, Drougge R. A new softening agent for melt spinning of softwood kraft lignin. J Appl Polym Sci 2013;129(3):1274–9. https://doi.org/10.1002/app.38795.

Otani S, Fukuoka Y, Igarashi B, Sasaki K. Method for producing carbonized lignin fibre. (Nippon Kayaku Kk), US Pat. 1969; 3461082.

Park S-J, Heo GY. Precursors and manufacturing of carbon fibers. In: Carbon fibers. vol. 210. Dordrecht, Netherlands: Springer; 2015. p. 31–66.

Sevastyanova O, Qin W, Kadla JF. Effect of nanofillers as reinforcement agents for lignin composite fibers. J Appl Polym Sci 2010;117(5):2877–81. https://doi.org/10.1002/app.32198.

Souto F, Calado V, Pereira Jr. N. Lignin-based carbon fiber: a current overview. Mater Res Express 2018;5:072001. https://doi.org/10.1088/2053-1591/aaba00.

Sudo K, Shimizu K. A new carbon fiber from lignin. J Appl Polym Sci 1992;44(1):127–34. https://doi.org/10.1002/app.1992.070440113.

Sudo K, Shimizu K, Nakashima N, Yokoyama A. A new modification method of exploded lignin for the preparation of a carbon fiber precursor. J Appl Polym Sci 1993;48(8):1485–91. https://doi.org/10.1002/app.1993.070480817.

Thunga M, Chen K, Grewell D, Kessler MR. Bio-renewable precursor fibers from lignin/polylactide blends for conversion to carbon fibers. Carbon 2014;68:159–66. https://doi.org/10.1016/j.carbon.2013.10.075.

Uraki Y, Nakatani A, Kubo S, Sano Y. Preparation of activated carbon fibers with large specific surface area from softwood acetic acid lignin. J Wood Sci 2001;47(6):465–9. https://doi.org/10.1007/BF00767899.

Wang G, Chen H. Fractionation of alkali-extracted lignin from steam-exploded stalk by gradient acid precipitation. Sep Purif Technol 2013;105:98–105. https://doi.org/10.1016/j.seppur.2012.12.009.

Wang S, Zhou Z, Xiang H, Chen W, Yin E, Chang T, Zhu M. Reinforcement of lignin-based carbon fibers with functionalized carbon nanotubes. Compos Sci Technol 2016;128:116–22. https://doi.org/10.1016/j.compscitech.2016.03.018.

Youe W-J, Lee S-M, Lee S-S, Lee S-H, Kim YS. Characterization of carbon nanofiber mats produced from electrospun lignin-g-polyacrylonitrile copolymer. Int J Biol Macromol 2016;82:497–504. https://doi.org/10.1016/j.ijbiomac.2015.10.022.

Zhang M. Carbon fibres derived from dry-spinning of modified lignin precursors [PhD thesis], Chemical Engineering, Clemson University: Clemson, SC, USA; 2016.

Zhang M, Ogale AA. Carbon fibers from dry-spinning of acetylated softwood kraft lignin. Carbon 2014;69:626–9. https://doi.org/10.1016/j.carbon.2013.12.015.

Further reading

Demuner IF, Colodette JL, Demuner AJ, Jardim CM. Biorefinery review: Wide-reaching products through kraft lignin. Bioresources 2019;14(3), 7543–7581.

CHAPTER 5

Properties of carbon fiber

Carbon fiber possesses a best combination of properties (Table 5.1). "The fibers do not suffer from stress corrosion or stress rupture failures at room temperatures, as glass and organic polymer fibers do. Especially at high temperatures, the strength and modulus are outstanding compared to other materials. Carbon fiber composites are ideally suited to applications where strength, stiffness, lower weight, and outstanding fatigue characteristics are critical requirements. They are also finding applications where high temperature, chemical inertness, and high damping are important. Carbon fibers also have good electrical conductivity, thermal conductivity, and low linear coefficient of thermal expansion" (Smith, 1987; Chand, 2000; Donnet and Bansal, 1990; polen.itu.edu.tr).

In Table 5.1, carbon fiber properties are compared with other materials (Hull and Clyne, 1996; Ventura and Martelli, 2009; Peebles, 1995; Bunsell and Renard, 2005). When comparison is made with steel, carbon fibers possess very high strength and also specific modulus. When compared with an absolute basis with Kevlar fibers, which are very strong, they show much higher modulus and strength. Overall, carbon fibers possess superb combination of strength, modulus, and conductivities in comparison with other materials.

Table 5.2 presents axial tensile properties of carbon fiber.

Carbon fiber can be classified into following two groups on the basis of its mechanical properties:
- general purpose
- high performance

The precursors used for producing carbon fibers are very important in determining the final properties and also its classification. Pitch obtained from petroleum or coal and polyacrylonitrile are the important precursors used for producing carbon fibers on a commercial scale. Virtually, 80% of commercial carbon fibers are produced from polyacrylonitrile due to its exceptional properties in comparison with those of carbon fiber obtained from pitch (Luo, 2010). Carbon fiber derived from polyacrylonitrile is costly, and because of this, its application to high performing materials is restricted. There is a necessity for a low-cost precursor that is able to produce carbon fiber with properties better in comparison with those of pitch and polyacrylonitrile. In commercial carbon fiber the carbon content should be no less than 92% carbon by weight.

Carbon fiber is available in various types. These are short or continuous. Structure can be partly crystalline or amorphous (Chung, 1994). Carbon layers tend to be parallel to the

Table 5.1 Comparison of carbon fibers properties with other materials.

Material	Specific gravity	Modulus (GPa)	Specific modulus (GPa)	Electrical resistivity (μΩm)	Thermal conductivity (W/mK)	Strength (GPa)	Specific strength (GPa)
Carbon fibers (high strength)	1.82	294	164	N/A	N/A	7.1	3.9
Kevlar fibers	1.45	135	93	N/A	0.04	2–3	~1.5–2
Steel	7.9	200	25	0.72	50	1–1.5	0.1–0.2
Glass (bulk and fiber)	2.5	72	28	10^{18}–10^{19}	13	2–3	~1–1.5
Aluminum	2.7	76	28	0.003	205	0.5	0.2

Based on Zhang M. Carbon fibres derived from dry-spinning of modified lignin precursors [Ph.D. thesis]. Clemson: Chemical Engineering, Clemson University; 2016; Peebles LH. Carbon fibres: structure and formation. New York, NY, USA: CRC Press; 1995. p. 12–168; Bunsell AR, Renard J. Fundamentals of fibre reinforced composite materials. London: IOP Publishing; 2005. p. 3–5; Hull D, Clyne T. An introduction to composite materials. Cambridge: Cambridge University Press; 1996; Ventura G, Martelli V. Thermal conductivity of Kevlar 49 between 7 and 290K. Cryogenics 2009;49:735–7.

Table 5.2 Axial tensile properties of carbon fibers.

	Tensile strength (GPa)	Tensile modulus (GPa)	Elongation at break (%)
Mesophase pitch	1.5–3.5	200–800	0.3–0.9
PAN	2.5–7.0	250–400	0.6–2.5
Rayon	~1.0	~50	~2.5

Based on Smith WS. Engineered materials handbook, vol. 1. Ohio: ASM International; 1987. p. 49.

fiber axis; therefore the modulus of elasticity is high. Fiber "texture" is a term described by the definition of the crystallographic direction aligned parallel to the fiber axis. "The modulus of elasticity of carbon fibre is higher parallel to the fibre axis than perpendicular to the axis. The stronger the fibre texture, the higher the degree of alignment of the carbon layer parallel to the fibre axis. Carbon fibre with high fibre texture has high strength and high tensile energy absorption. Carbon fibre generally has excellent tensile strength, low densities, high thermal and chemical stabilities in the absence of oxidizing agents, good thermal and electrical conductivities and excellent creep resistance. Carbon fibre generally comes in the form of woven textile, prepreg, continuous fibres, rovings and chopped fibres. To further process the composite parts into final product, the composite parts can be produced through filament winding, tape winding, pultrusion, compression moulding, vacuum bagging, liquid moulding or injection moulding" (Bajpai, 2017; Huang, 2009).

Carbon fiber can be grouped into several categories based on its modulus of elasticity (Table 5.3). Carbon fibers are classified as low modulus, standard modulus, intermediate modulus, high modulus, and ultrahigh modulus. "Low modulus have a tensile modulus below 34.8 million psi (240 million kPa). Ultra-high modulus carbon fibres have a tensile modulus of 72.5–145.0 million psi (500 million–1.0 billion kPa). As a comparison, steel has a tensile modulus of about 29 million psi (200 million kPa). Thus, the strongest carbon fibres are ten times stronger than steel and eight times that of aluminium, not to mention much lighter than both materials, 5 and 1.5 times, respectively. Additionally, their fatigue properties are superior to all known metallic structures, and they are one of the most corrosion-resistant materials available, when coupled with the proper resins" (Price, 2011).

Prices are also found to vary significantly among different categories (Table 5.3). High- and ultrahigh-stiffness fiber are very expensive (2000 per kg) and are used in special applications such as airfoils. The cost of standard modulus fiber is around $22 per kg. These are used in civil infrastructure industry.

Carbon fibers are the strongest and the stiffest reinforcing fibers for polymer composites, the most used after glass fibers. These fibers have low density and a negative coefficient of longitudinal thermal expansion (Chung, 1994; Donnet and Bansal, 1990; Minus and Kumar, 2005, 2007; Xiaosong, 2009). The tensile energy absorption refers to energy

Table 5.3 Prices of carbon fiber of different categories.

Low modulus
Modulus of elasticity—40 to 200 GPa
Less than $20/kg
Nonstructure usage
Standard modulus
Modulus of elasticity—200 to 275 GPa
$20–$55/kg
Automotive, sporting goods, wind turbine, pressure tanks
Intermediate modulus
Modulus of elasticity—275 to 345 GPa
$55–$65/kg
Pressure tanks, wind turbine
High modulus
Modulus of elasticity—345 to 600 GPa
$65–$90/kg
Aviation, military
Ultrahigh modulus
Modulus of elasticity—600 to 965 GPa
Up to $2000/kg
Aerospace, military

Based on Chen MCW. Commercial viability analysis of lignin based carbon fibre [Master thesis]. Simon Fraser University; 2014; Eberle C. Carbon fibre from lignin. Oak Ridge, TN. Retrieved from: https://www.cfcomposites.org/PDF/Breakout_Cliff.pdf; 2012; Price R.E. Carbon fibre used in fibre reinforced plastic (FRP). https://www.build-on-prince.com/carbon-fibre.html; 2011.

stored in the fiber when the fiber is under tension with force and undergoes all extension or changes in length.

Carbon fibers are very expensive and can give galvanic corrosion in contact with metals (Chung, 1994; Donnet and Bansal, 1990; Xiaosong, 2009). They are generally used together with epoxy, where high strength and stiffness are required, that is, race cars, automotive and space applications, and sport equipment. Depending on the orientation of the fiber, the carbon fiber composite can be stronger in a certain direction or equally strong in all directions. A small piece can withstand an impact of many tons and still deform minimally. The properties of a carbon fiber part are close to that of steel, and the weight is close to that of plastic. Thus the strength-to-weight ratio and stiffness-to-weight ratio of a carbon fiber part are much higher than either steel or plastic. The specific details depend on the matter of construction of the part and the application. For instance, a foam core sandwich has extremely high strength-to-weight ratio in bending, but not necessarily in compression or crush. In addition, the loading and boundary conditions for any components are unique to the structure within which they reside. The modulus of carbon fiber is typically 20 msi, and its ultimate tensile strength is typically 500 ksi. High-stiffness and high-strength carbon fiber materials are also available through specialized heat treatment processes with much higher values, in comparison with 2024-T3 aluminum, which has a modulus of only 10 msi and ultimate tensile strength of 65 ksi, and 4130 steel, which has a modulus of 30 msi and ultimate tensile strength of 125 ksi.

Carbon fibers can be classified based on carbon fiber properties, precursor fiber materials, and final heat treatment temperature (Table 5.4). Table 5.5 shows mechanical properties of different types of carbon fibers. Table 5.6 shows mechanical properties of cellulosic fibers.

Table 5.7 shows properties of carbon fibers from different precursors.

Most common uses for carbon fiber are in applications where high strength to weight and high stiffness to weight are desirable (Chung, 1994; Donnet and Bansal, 1990; Minus and Kumar, 2005). These include aerospace, military structures, sports equipment, robotics, wind turbines, and manufacturing fixtures. High toughness can be achieved by combining with other materials. Certain applications also use carbon fiber's electrical

Table 5.4 Classification of carbon fiber.

Based on carbon fiber properties

- Ultrahigh modulus 600–965 GPa
- High modulus 345–600 GPa
- Intermediate modulus 275–345 GPa
- Standard modulus 200–275 GPa
- Low modulus 40–200 GPa
- High tensile >3.0 GPa
- Superhigh tensile >4.5 GPa

Based on precursor fiber materials

- PAN-based carbon fibers
- Pitch-based carbon fibers
- Mesophase pitch-based carbon fibers
- Isotropic pitch-based carbon fibers
- Rayon-based carbon fibers
- Gas phase-grown carbon fibers

Based on final heat treatment temperature

- High-heat treatment carbon fibers (HTT), where final heat treatment temperature should be above 2000°C and can be associated with high-modulus type fiber
- Intermediate-heat treatment carbon fibers (IHT), where final heat treatment temperature should be around or above 1500°C and can be associated with high-strength type fiber
- Low-heat treatment carbon fibers, where final heat treatment temperatures not greater than 1000°C. These are low-modulus and low-strength materials

Based on Donnet JB, Bansal R.C. Carbon fibres. 2nd ed. New York: Marcel Dekker; 1990. p. 1–145; Minus ML, Kumar S. The processing, properties, and structure of carbon fibers. JOM 2005;57(2):52–8.

Table 5.5 Mechanical properties of different types of carbon fibers.

Type of fibers	Tensile strength (GPa)	Young modulus (GPa)
High strength—HT	3.3–6.9	200–250
Intermediate modulus—IM	4.0–5.8	280–300
High modulus—HM	3.8–4.5	350–600
Ultrahigh modulus—UHM	2.4–3.8	600–960

Based on Donnet JB, Bansal RC. Carbon fibres. 2nd ed. New York: Marcel Dekker; 1990. p. 1–145; Minus ML, Kumar S. The processing, properties, and structure of carbon fibers. JOM 2005;57(2):52–8; Xiaosong H. Fabrication and properties of carbon fibers. Materials 2009;2:2369–403.

Table 5.6 Mechanical properties of cellulosic fibers.

Sample	Sample linear density (tex)	Specific strength (N/tex)	Specific stress (N/tex)	Elongation at break (%)	DP	Crystallinity %
Viscose	0.167	0.42	3.01	16	250–350	45–48
Tencel	0.130	0.40		13		
Lyocell	0.130	0.61	7.44	6.4	550–600	70–75
Bacterial cellulose (BC)			20–60			60–80
Regenerated BC fibers		0.05–0.15		3–8		

Based on Dumanlı AG, Windle AH. Carbon fibres from cellulosic precursors: a review. J Mater Sci 2012;47(10):4236–50; Röder T, Moosbauer J, Kliba G, Schlader S, Zuckerstätter G, Sixta H. Comparative characterization of man-made regenerated cellulose fibres. Lenzinger Ber 2009;87:98–105; Sfiligoj Smole M, Persin Z, Kreze T, Stana Kleinschek K, Ribitsch V, Neumayer S. X-ray study of pre-treated regenerated cellulose fibres. Mater Res Innov 2003;7:275; Guhados G, Wan W, Hutter JL. Measurement of the elastic modulus of single bacterial cellulose fibers using atomic force microscopy. Langmuir 2005;21:6642–6; Watanabe K, Tabuchi M, Morinaga Y, Yoshinaga F. Structural features and properties of bacterial cellulose produced in agitated culture. Cellulose 1998;5:187–200; Yamanaka S. J Mater Sci 1989;24:3141. https://doi.org/10.1007/BF01139032; Nishi Y, Uryu M, Yamanaka S, Watanabe K, Kitamura N, Iguchi M, Mitsuhashi S. The structure and mechanical properties of sheets prepared from bacterial cellulose. J Mater Sci 1990;25:2997–3001.

conductivity and also high thermal conductivity in the case of certain special applications. Carbon fiber produces a very good surface finish in addition to the basic mechanical properties. Despite the fact that carbon fiber shows significant advantages in comparison with other materials, there are also trade-offs one should weigh against. First, solid carbon fiber does not undergo plastic deformation. Underload carbon fibers bend but do not get deformed forever. But once the ultimate strength of the material exceeds, carbon fiber fails suddenly and disastrously. In designing the process, it is crucial that the engineer understands and accounts for this behavior, especially the design safety factors. "Carbon

Table 5.7 Properties of carbon fibers from different precursors.

Carbon fiber	Density (g/cm^3)	Tensile strength (GPa)	Tensile modulus (Gpa)	Elongation at break (%)	Diameter (lm)
PAN-based CF	1.70–1.80	3.5–6.3	200–500	0.8–2.2	5–10
Pitch-based CF	1.80–2.20	1.3–3.1	150–900	0.3–0.9	10–11
Rayon-based CF	1.40–1.50	0.5–1.2	40–100		5–10
Lyocell-based CF		0.9–1.1	90–100	1.0–1.1	8

Based on Dumanlı AG, Windle AH. Carbon fibres from cellulosic precursors: a review. J Mater Sci 2012;47(10):4236–50; Peng S, Shao H, Hu X. Lyocell fibers as the precursor of carbon fibers. J Appl Polym Sci 2003;90:1941–7; Zhang H, Guo L, Shao H, Hu X. Nano-carbon black filled lyocell fiber as a precursor for carbon fiber. J Appl Polym Sci 2006;99:65–74; https://www.toraycfa.com/product.html; https://cytec.com/engineered-materials/thornel-pitch.htm; https://www.sohim.by/en/catalog/carbon/.

fibre composites are also very expensive as compared to traditional materials. Working with carbon fibre needs a high skill and several complicated processes for producing high quality building materials (for instance, sandwich laminates, solid carbon sheets, tubes, etc). Very high skill and specialized tooling and machinery are needed for creating custom-fabricated, highly optimized parts and assemblies" (dragonplate.com).

When designing composite parts, one cannot simply compare properties of carbon fiber versus steel, aluminum, or plastic, since these materials are in general homogeneous (properties are the same at all points in the part) and have isotropic properties throughout (properties are the same along all axes). By comparison, in a carbon fiber part, the strength resides along the axis of the fibers, and thus fiber properties and orientation greatly impact mechanical properties (Peebles, 1994, 1995; Johnson and Watt, 1967; Wicks, 1975; Watt and Johnson, 1969; Fourdeux et al., 1971; Perret and Ruland, 1970; Diefendorf and Tokarsky, 1975; Edie, 1998; Endo, 1988). Carbon fiber parts are in general neither homogeneous nor isotropic (Chung, 1994; Donnet and Bansal, 1990).

Carbon fibers typically exhibit a skin-core texture that has been confirmed using optical microscopy (Chung, 1994; Minus and Kumar, 2005). The skin can result from higher preferred orientation and a higher density of material at the fiber surface. The formation of the skin is also associated with the coagulation conditions during PAN precursor fiber spinning. The fine structure of carbon fibers consists of basic structural units of turbostratic carbon planes. The distance between turbostratic planes and perfect graphite planes is generally >0.34 nm and 0.3345 nm, respectively (Hoffman et al., 1991; Johnson, 1979).

Typical structural parameters for the selected pitch and polyacrylonitrile-based carbon fibers are given in Table 5.8. The crystallite size in the high-modulus pitch-based fibers is as high as 25 nm along the c-axis direction and is 64 nm along the a-axis parallel to the fiber

Table 5.8 Typical structural parameters for the selected pitch- and PAN-based carbon fibers.

Structural parameters	P-25	P-55	P-100	P-120	T-300	IM-8
Crystal size parallel to c-axis	2.6	12.4	22.7	25.1	1.5	8
Crystal size parallel to a-axis and perpendicular to the fiber axis direction	4	11	49	64	2.2	1.9
Crystal size parallel to a-axis and parallel to the fiber axis direction	6	30	80	88	4.1	3.1
Orientation parameter, full width at half maximum of the (002) azimuthal scan in degrees Fi	31.9	14.1	5.6	5.6	35.1	5.1
d(002)	0.344	0.342	0.3382	0.3376	0.342	0.343
SEM morphology	Sheet like	Sheet like	Sheet like	Sheet like	No	No

Based on Minus ML, Kumar S. The processing, properties, and structure of carbon fibers. JOM 2005;57(2):52–8; Minus ML, Kumar S. Carbon fibre. Kirk-Othmer Encycl Chem Technol 2007;26:729–49.

axis and 88 nm along the a-axis perpendicular to the fiber axis. Crystallite dimensions in fibers such as K-1100 are expected to be even larger.

"The crystallite size in the PAN-based carbon fibres (T-300 and IM-8) is in the 1.5- to 5-nm range. High-modulus pitch-based carbon fibres exhibit high orientation (Z ¼ 5.6), whereas the orientation of the pitch-based carbon fibres is relatively low (Z ¼ 35.1). High-modulus pitch-based carbon fibres (P-100 and P-120) also exhibit graphitic sheet-like morphology from scanning electron microscopy and, as well as clear evidence of the three-dimensional order from X-ray diffraction" (Kumar et al., 1993). Because of the formation of microdomains, which can bend and twist, carbon fiber contains defects, vacancies, dislocations, grain boundaries, and impurities. Low interlayer spacing, large crystallite size, high degree of orientation parallel to the fiber axis, low density of defects, and high crystallinity are characteristics of the high tensile modulus and high thermal and electrical conductivity fibers. Porosity in carbon fibers is measured using small-angle X-ray scattering, and these data can be used for estimating the size, shape, and orientation of the pores. Pore size and distribution and pore orientation change as the fiber goes through increasing heat treatment and tension (Chwastiak, 1980; Peterlik and Kromp, 1994; info.smithersrapra.com).

Carbon fiber properties are related to the fiber morphology and microstructure. Properties of some commercial carbon fibers are presented in Table 5.9.

The axial compressive strength of polyacrylonitrile-based carbon fibers is higher than those of the pitch-based fibers, and it decreases with increasing modulus in both cases. It is understood that higher orientation, higher graphitic order, and larger crystal size all contribute negatively to the compressive strength. Polyacrylonitrile-based carbon fibers typically fail in the buckling mode, whereas pitch-based fibers fail by shearing mechanisms

Table 5.9 Properties of commercial carbon fibers.

	Tensile strength (GPa)	Tensile modulus (GPa)	Elongation to break (%)	Density, r (g/cm^3)	Thermal conductivity (W/mK)	Electrical conductivity (S/m)
Toray Torayca1 PAN based						
T300	3.53	230	1.51	1.76		
T700SC	4.90	230	2.1	1.80		
M35JB	4.70	343	1.4	1.75		
M50JB	4.12	475	0.9	1.88		
M55J	4.02	540	0.8	1.91		
M30SC	5.49	294	1.9	1.73		
Cytec Thomel1 PAN based						
T300	3.75	231	1.4	1.76	8	5.56E+04
T650/35	4.28	255	1.7	1.77	14	6.67E+04
T300C	3.75	231	1.4	1.76	8	5.56E+04

Based on Donnet JB, Bansal R.C. Carbon fibres. 2nd ed. New York: Marcel Dekker; 1990. p. 1–145; Minus ML, Kumar S. The processing, properties, and structure of carbon fibers. JOM 2005;57(2):52–8.

(Dobb et al., 1990). This suggests that the compressive strength of intermediate modulus polyacrylonitrile-based carbon fibers may be higher than what is being realized in the composites. Changes in the fiber geometry, effective fiber aspect ratio, fiber/matrix interfacial strength, and also matrix stiffness can result in fiber compressive strength increase, until the failure mode changes from buckling to shear. High compressive strength fibers also exhibit high shear modulus (Kumar et al., 1993). Compressive strength dependence of pitch- and polyacrylonitrile-based carbon fibers on several structural parameters has been studied (Kumar et al., 1993). Kozey et al. (1995) have reviewed the compressive strength of high-performance fibers and also compression test methods. The electrical and thermal conductivities increase with increasing fiber modulus and carbonization temperature (Issi and Nysten, 1998; https://www.cytec.com; https://www.toray.com).

> *The electrical conductivity of polyacrylonitrile-based carbon fibres is in the range of 104 to 105 S/m, and that of the pitch-based carbon fibres is in the range of 105 to 106 S/m. The electrical conductivity increases with temperature because as the temperature is raised, the density and carrier (electrons and holes) mobility increases. Defects are known to cause carrier scattering. An increase in modulus is due to increased orientation of the carbon planes; this decreases the concentration of defects and subsequently decreases carrier scattering. The thermal conductivity of pitch-based carbon fibres is in the range of 20–1000 W/mK. Carbon fibre resistance to oxidation increases with the degree of graphitization.*
>
> *(Minus and Kumar, 2007)*

For carbon fibres, thermal gravimetric analysis in air shows the initial weight loss above 400°C, sharp weight loss in the 500–600°C range, and total weight loss by 850°C. Axial coefficient of thermal expansion of the 200- to 300-GPa modulus carbon fibres is in the range of 0.4 to 0.8×10^6/C. For the high-modulus (700 to 900 GPa) carbon fibres, it is about 1.6×10^6/C.

(Minus and Kumar, 2005, 2007)

Compressive properties dictate the use of carbon composites in several structural applications. Extensive research has been conducted on compressive properties and morphology of carbon fibers. Increased oxidation resistance at high temperatures is also one of the important requirements for some specialized uses (Chand, 2000).

Carbon fiber-reinforced composites can be used in the design of advanced materials and systems. "The properties of the fibre-reinforced plastic articles are governed mainly by the properties of the fibre, in particular the carbon fibre, and the form of textile into which the fibre is processed. Preimpregnated materials (prepregs) offer a precise and economical way of combining reinforcements with a resin matrix. Prepregs consist of high-quality textile fabrics impregnated with curable resins. The fibre type is the main factor governing the strength, Young's modulus, and other important properties of fibre composite products. High strength, rigidity and pronounced anisotropy are achieved by a unidirectional (UD) arrangement of the fibres or the prepregs themselves" (www.carbonandgraphite.org) (Table 5.10).

As the fibres are arranged in dense bundles, the unidirectional prepregs contain at least 60 percent fibres by volume. In principle, prepregs made from woven fabrics are employed for components that have to be isotropic in one plane (orthotropic). This can be achieved with plain-weave fabrics, in which warp and weft are arranged at angles of +45°/−45° and 0°/90° to the main axis of the laminate. In general, the fibre content of such elements will be about 50 percent by volume. Not only does the resin influence the essential properties of the resulting products, but it also determines their processibility, manufacturing time.

(www.carbonandgraphite.org)

Carbon nanotubes have extraordinary mechanical, electrical, and thermal properties (Treacy et al., 1996; Thess et al., 1996; Dresselhaus and Eklund, 2000; Dresselhaus et al., 1995; Iijima and Ichihashi, 1993; Nikolaev et al., 1999). Single-wall nanotubes can be thought of as the ultimate carbon fiber due to their perfect graphitic structure, low density, and alignment with respect to each layer that gives them exceptional engineering properties and lightweight. The elastic modulus parallel to the nanotubes axis is estimated to be ~640 GPa and the tensile strength to be ~37 GPa (Gao et al., 1998; Walters et al., 1999). Single-wall nanotube electrical and thermal conductivity at 300 K are 106 S/m (Berber et al., 2000) and ~3000 W/mK (Kim et al., 2001), respectively. The combination of density, mechanical, thermal, and electrical properties of single-wall

Table 5.10 Advantages of carbon fiber-reinforced carbon composites.

Resistance to high temperatures and weathering, low flammability, low smoke density, and low toxicity of decomposition products. Temperature resistance of course depends on choice of resin
High chemical stability
Large variety of possible component shapes and sizes
High durability due to long prepreg storage life
Prepregs comprise the range of reinforcements and resin matrix combinations. They are manufactured on a state-of-the-art fusible resin plant. Fusible resins have fewer volatile constituents and increase the composite materials' mechanical strength. The prepreg manufacturing plant is accredited to DIN AND ISO 9001 quality assurance standards

Based on Donnet JB, Bansal R.C. Carbon fibres. 2nd ed. New York: Marcel Dekker; 1990. p. 1–145; Minus ML, Kumar S. The processing, properties, and structure of carbon fibers. JOM 2005;57(2):52–8; Minus ML, Kumar S. Carbon fibre. Kirk-Othmer Encycl Chem Technol 2007;26:729–49.

nanotubes is unmatched, as there are no other materials with this combination of properties. The translation of these properties into macroscopic structures is the subject of current challenge for the material scientists and engineers. The intrinsic mechanical and transport properties of carbon nanotubes make them the ultimate carbon fibers (https://www.nanocyl.com/en/CNT-Expertise-Centre/Carbon-Nanotubes).

Tables 5.11 and 5.12 compare these properties with other engineering materials. Overall, carbon nanotubes show a unique combination of stiffness, strength, and tenacity compared with other fiber materials that usually lack one or more of these properties.

Table 5.11 Properties of various engineering fibers.

Fiber material	Specific density	E (TPa)	Strength (GPa)	Strain at break (%)
Carbon nanotube	1.3–2	1	10–60	10
HS steel	7.8	0.2	4.1	< 10
Carbon fiber—PAN	1.7–2	0.2–0.6	1.7–5	0.3–2.4
Carbon fiber—pitch	2–2.2	0.4–0.96	2.2–3.3	0.27–0.6
E/S—glass	2.5	0.07/0.08	2.4/4.5	4.8
Kevlar 49	1.4	0.13	3.6–4.1	2.8

Based on https://www.nanocyl.com/en/CNT-Expertise-Centre/Carbon-Nanotubes.

Table 5.12 Thermal conductivity and electrical conductivity of carbon fiber and carbon nanotubes.

Fiber material	Thermal conductivity (W/m k)	Electrical conductivity
Carbon nanotubes	> 3000	106–107
Carbon fiber—pitch	1000	2–8.5 × 106
Carbon fiber—PAN	8–105	6.5–14 × 106

Based on https://www.nanocyl.com/en/CNT-Expertise-Centre/Carbon-Nanotubes.

Thermal and electrical conductivity are also very high, and comparable with other conductive materials. The properties of nanotubes have caused researchers and companies to consider using them in several fields. For example, because carbon nanotubes have the highest strength-to-weight ratio of any known material, researchers at NASA are combining carbon nanotubes with other materials into composites that can be used to build lightweight spacecraft.

Another property of nanotubes is that they can easily penetrate membranes such as cell walls. In fact, nanotube's long, narrow shape make them look like miniature needles, so it makes sense that they can function like a needle at the cellular level. Medical researchers are using this property by attaching molecules that are attracted to cancer cells to nanotubes to deliver drugs directly to diseased cells.

Another interesting property of carbon nanotubes is that their electrical resistance changes significantly when other molecules attach themselves to the carbon atoms. Companies are using this property to develop sensors that can detect chemical vapors such as carbon monoxide or biological molecules. Researchers and companies are working to use carbon nanotubes in various fields.

References

Bajpai P. Carbon fibre from lignin. In: SpringerBriefs in material science. Springer (Springer Nature); 2017.
Berber S, Kwon YK, Tomanek D. Unusually high thermal conductivity of carbon nanotubes. Phys Rev Lett 2000;84(20):4613–6.
Bunsell AR, Renard J. Fundamentals of fibre reinforced composite materials. London: IOP Publishing; 20053–5.
Chand SJ. Carbon fibers for composites. Mater Sci 2000;35:1303–13.
Chung DL. Carbon fibre composites. Boston: Butterworth-Heinemann; 19943–11.
Chwastiak S. Inventor; Union Carbide Corporation, assignee; US4,032,430 1980.
Diefendorf RJ, Tokarsky E. High performance carbon fibers. Polym Eng Sci 1975;15:150.
Dobb MG, Johnson DJ, Park CR. Compression behavior of carbon fibers. J Mater Sci 1990;25(7):829–34.
Donnet JB, Bansal RC. Carbon fibres. 2nd ed. New York: Marcel Dekker; 19901–145.
Dresselhaus MS, Eklund PC. Phonons in carbon nanotubes. Adv Phys 2000;49:705–814.
Dresselhaus MS, Dresselhaus G, Saito R. Physics of carbon nanotubes. Carbon 1995;33(7):883–91.
Edie DD. The effect of processing on the structure and properties of carbon fibers. Carbon 1998;36(4):345–62.
Endo M. Structure of mesophase pitch-based carbon fibers. J Mater Sci 1988;23:598–605.
Fourdeux A, Perret R, Ruland W. Carbon fibres: their composites and applications. London: The Plastics Institute; 1971, p. 57.
Gao GH, Cagin T, Goddard WA. Energetics, structure, mechanical and vibrational properties of single-walled carbon nanotubes. Nanotechnology 1998;9(3):184–91.
Hoffman WP, Hurley WC, Liu PM, Owens TW. The surface topography of nonshear treated pitch and PAN carbon fibers as viewed by the STM. J Mater Res 1991;6(8):1685–94.
Hull D, Clyne T. An introduction to composite materials. Cambridge: Cambridge University Press; 1996.
Huang X. Fabrication and properties of carbon fibres. Materials 2009. 2:2369–2403. https://doi.org/10.3390/ma2042369.
Iijima S, Ichihashi T. Single-shell carbon nanotubes of 1-nm diameter. Nature 1993;363:603–5.
Issi JP, Nysten B. Electrical and thermal transport properties in carbon fibers. In: Donnet J-B, Rebouillat S, Wang TK, Peng JCM, editors. Carbon fibers. New York: Marcel Dekker; 1998. p. 371–461.
Johnson DJ. Nature 1979;279(10):142.

Johnson W, Watt W. Structure of high modulus carbon fibres. Nature 1967;215(5099):384–6.

Kim P, Shi L, Majumdar A, McEuen PL. Thermal transport measurements of individual multiwalled nanotubes. Phys Rev Lett 2001;87(21):215502.

Kozey VV, Jiang H, Mehta VR, Kumar S. Compressive behavior of materials: Part II, high performance fibers. J Mater Res 1995;10:1044–61.

Kumar S, Anderson DP, Crasto AS. Carbon fibre compressive strength and its dependence on structure and morphology. J Mater Sci 1993;28(2):423–39.

Luo J. Lignin-based carbon fibre. [A thesis, Degree of Master of Science in Chemical Engineering], Maine: The University of Maine; 2010.

Minus ML, Kumar S. The processing, properties, and structure of carbon fibers. JOM 2005;57(2):52–8.

Minus ML, Kumar S. Carbon fibre. Kirk-Othmer Encycl Chem Technol 2007;26:729–49.

Nikolaev P, Bronikowski MJ, Bradley RK, Rohmund F, Colbert DT, Smith KA, Smalley RE. Gas-phase catalytic growth of single-walled carbon nanotubes from carbon monoxide. Chem Phys Lett 1999;313(1–2):91–7.

Peebles LH. Carbon fibers: formation, structure, and properties. 1st ed. Boca Raton, FL, USA: CRC Press; 1994, p. 3.

Peebles LH. Carbon fibres: structure and formation. New York, NY, USA: CRC Press; 199512–168.

Perret R, Ruland W. The microstructure of PAN based carbon fibers. J Appl Cryst 1970;3:525.

Peterlik FP, Kromp K. Pore structure of carbon–carbon composites studied by small-angle X-ray scattering. Carbon 1994;32(5):939–45.

Price RE. Carbon fibre used in fibre reinforced plastic (FRP), https://www.build-on-prince.com/carbon-fibre.html; 2011.

Smith WS. Engineered materials handbook. vol. 1. Ohio: ASM International; 1987, p. 49.

Thess A, Lee R, Nikolaev P, Dai H, Petit P, Robert J, Xu C, Lee YH, Kim SG, Rinzler AG, Colbert DT, Scuseria GE, Tomanek D, Fischer JE, Smalley RE. Crystalline ropes of metallic carbon nanotubes. Science 1996;273(5274):483–7.

Treacy M, Ebbesen TW, Gibson JM. Exceptionally high Young's modulus observed for individual carbon nanotubes. Nature 1996;381(6584):678–80.

Ventura G, Martelli V. Thermal conductivity of Kevlar 49 between 7 and 290K. Cryogenics 2009;49:735–7.

Walters DA, Ericson LM, Casavant MJ, Liu J, Colbert DT, Smith KA, Smalley RE. Elastic strain of freely suspended single-wall carbon nanotube ropes. Appl Phys Lett 1999;74(25):3803–5.

Watt W, Johnson W. High temperature resistant fibers from organic polymer. Appl Polymer Symp 1969;9:229–43.

Wicks BJ. Microstructural disorder and the mechanical properties of carbon fibers. J Nucl Mater 1975;56(3):287–96.

Xiaosong H. Fabrication and properties of carbon fibers. Materials 2009;2:2369–403.

Further reading

Association Khimvolokno. https://www.sohim.by/en/catalog/carbon/.

Carbon nanotubes, https://www.nanocyl.com/en/CNT-Expertise-Centre/Carbon-Nanotubes.

Cytec Industries. https://www.cytec.com.

CYTEC.COM. https://cytec.com/engineered-materials/thornel-pitch.htm; 2009.

Fitzer E. Figueiredo JL, Bernardo CA, RTK B, Huttinger KJ, editors. Carbon fibres filaments and composites. Dordrecht: Kluwer Academic; 1990 p. 3–4, 43–72, 119–146.

Gao Q, Shen X, Lu X. Regenerated bacterial cellulose fibers prepared by NMMO·H2O process. Carbohydr Polym 2011;83:1253–6.

SOHIM. Republican Unitary Enterprise Svetlogorsk Production Association Khimvolokno. 2010. https://www.sohim.gomel.by.

Toray Carbon Fibers America Inc. https://www.toraycfa.com/product.html; 2010.

Toray Global. https://www.toray.com.

Watt W. Kelly A, Rabotnov YN, editors. Handbook of composites. vol. I. Holland: Elsevier Science; 1985. p. 327–87.

CHAPTER 6

Recycling of carbon fiber-reinforced polymers

Carbon fiber-reinforced polymers (CFRPs) are extremely advantageous materials in lightweight structural applications because of their high strength-to-weight ratio and resistance to corrosion (Xu et al., 2013). These materials are used in wind turbine, sports industries aerospace, and automobile. CFRPs are especially useful in aircraft applications mainly due to the material's weight-saving properties. CFRP use also contributes to large-scale reductions in fuel consumption and greenhouse gas emissions. New aircrafts such as the Boeing 787 and the Airbus A350 contain more than 50% carbon composite materials by weight (Pimenta and Pinho, 2011). The advantageous properties of CFRPs have led to a large increase in its use (Maxime, 2014). During the last decade the global demand for CFRPs has risen substantially (Witik et al., 2013).

> *In 2017, the worldwide demand for Carbon Fibers was 82 400 tons, and this demand is expected to reach 112 000 tons in 2020. Since the service life of CFRPs is approximately 50 years, the extensive use of CFRP is now starting to produce serious waste disposal problems.*
>
> *(Zhu et al., 2019)*

CFRP is expected to be valued at over 25 billion dollars per year by 2025 (Roberts, 2006). But the increasing use of CFRP has led to an increasing amount of waste being generated, consisting primarily of scraps and end-of-life components (Jiang et al., 2008). For instance, increasingly more plants using CFRPs will end their service life, generating massive amounts of CFRP waste (Sun et al., 2015; Marsh, 2008; McConnell, 2010; Roberts, 2007).

Thus the treatment of CFRP waste is becoming a more urgent problem.

As a result of increased CFRP use, there has also been a rise in carbon composite waste. Manufacturing waste accounts for approximately 40% of all CFRP waste, while the remaining 60% accounts for components at end of life (EoL). Notably, by 2025, 8500 commercial aircrafts will be decommissioned, representing more than 20 tonnes of CFRP waste per aircraft (Pimenta and Pinho, 2011). Unfortunately today, most CFRP waste is landfilled due to technical and economical complexities of its recycling processes. Landfilling and waste incineration are extremely detrimental to the environment and promote disposal of nonrenewable materials. As it happens the European Union requires that 85% of a vehicle must be reusable or recyclable (Morin et al., 2012). With increasing environmental regulations, it is important to find reliable CFRP recycling techniques for preserving its use.

"A small fraction of the carbon fibre composite materials used is recycled. However, new legislation polices and approaching shortage of raw materials, in combination with the ever-increasing use of carbon fibre composites, will force society to recycle these materials in the near future. Until recently, the growing number of composite components from retired aircraft, wind turbine blades and automobiles have been disposed in landfills or incinerated. Even though landfills are the least expensive way of disposal they are not always an option. Landfill and incineration account for >90% of the disposal methods for CF structures. Since a European Union directive came into force in 2004, many member states are forbidding landfill disposal of composites. Also, new automotive legislation forces increased reuse of materials. The End-of-life (EoL) Vehicle Directive (EU 2000/53/EC, 2000), issued in year 2000, requires that 95 wt% of a vehicle manufactured after January 2015 has to be reused or recovered. The directive assigns the original equipment manufacturer to design recyclable products" (Pimenta and Pinho, 2011; Pickering, 2006). Recycled fiber consumes approximately 1/10 of the energy needed for producing virgin fiber. Table 6.1 shows the advantages of recycled carbon fiber (RCF).

There is a high potential for reuse of carbon fibre because demand for chopped and milled carbon fibre is growing with the use of carbon fibre outside the aerospace market. For carbon fibre composites, solving the waste stream is complicated due to the complex nature of the composites.

(pure.ltu.se)

Table 6.1 Advantages of recycled carbon fibers.

Excellent products
After the pyrolysis process the fibers have a highly active "raw" surface offering excellent bonding characteristics
The fibers are highly conductive

Economic gains
Lower-cost, high-value products
Stable pricing—not affected by worldwide carbon prices, so we can hold prices stable on a long-term basis
Stable supply—not be affected by any carbon shortages when supply "tightens"

Social responsibility
The recycling process uses approximately 1/10th of the energy required to produce virgin carbon fiber
Minimize waste going to landfill

Global demand for carbon fiber is forecast to grow to 140,000 tonnes by 2020. CFRP is used in various applications (Marsh, 2008; Sloan, 2007). Despite the benefits associated with CFRP, their increasing use also produces an increasing amount of CFRP waste. Common sources of waste are presented in the succeeding text:
- out-of-date prepregs,
- manufacturing cutoffs,
- testing materials,
- production tools,
- EoL components.

According to Pickering (2006), manufacturing waste is about 40% of all the CFRP waste produced, whereas woven trimmings contribute more than 60%. Recycling composites is difficult due to their complex nature (they contain fibers, matrix, and fillers), the cross-linked nature of thermoset resins that cannot be remolded, and their combination with other materials such as metal fixings, honeycombs, and hybrid composites. "Converting CFRP waste into a valuable resource and closing the loop in the CFRP lifecycle is important for the continued use of the material in certain applications (e.g., the automotive industry. This has motivated a considerable amount of research on recycling processes for CFRP over the last two decades" (cordis.europa.eu; Pickering, 2006).

Currently, most of the CFRP waste is landfilled. The airframe of EoL vehicles is typically disposed in desert graveyards, airports, or by landfilling (Pickering, 2006; Carberry, 2008; PAMELA, 2008). But these are unacceptable solutions because the increasing amount of CFRP produced raises serious issues about waste disposal and the consumption of nonrenewable resources (Pimenta and Pinho, 2011, 2014). Moreover, current European legislation is enforcing strict control of composite disposal. The responsibility for disposing of end-of-life (EoL) composites is now the manufacturer of the component. Legal landfilling of CFRP is not much. Automotive vehicles disposed after 2015 must be 85% recyclable. The production cost of carbon fiber is very high due to high energy consumption during manufacturing and the price of materials (Carberry, 2008). Demand for virgin carbon fiber generally surpasses supply capacity; therefore RCF could be reintroduced into the market for noncritical applications (Carberry, 2009). Disposal of CFRP by landfilling (if legal) can cost about £0.20 per kg (Meyer et al., 2007). Hence, recycling can convert expensive waste disposal into profitable reusable material. Carbon fiber and prepregs can be recycled, and the resulting recyclate retains more than 90% of the mechanical properties of the fiber.

With recent announcements by recycled carbon fiber, Limited (West Midlands, United Kingdom) and Materials Innovation Technologies (Fletcher, North Carolina, United States) to open recycling facilities for carbon fiber composites for commercial operation, RCF could become readily available. Several research institutions and companies have been involved in developing the methods for recycling carbon fiber. Carbon

Table 6.2 Carbon fiber recycling industry progress.

Recycled Carbon Fibre Limited, United Kingdom has gone into full-scale production
Focusing on milled carbon, now part of ELG
Materials Innovation Technologies, United States scaled up to production scale process
Vertically integrating with wetly processing, good match for rCF material form
RCF and MIT both actively collecting and processing trim scrap from Boeing carbon fiber composites supply chain
JCMA/MITI Japan conducting pilot scale trials
CFK Valley Stade announced plans for development of full-scale process

fiber recycling has now become a reality. The progress of the carbon fiber recycling industry is shown in Table 6.2.

> Carbon Fibre recycling offers three shades of green. It not only prevents the waste of virgin carbon fibre in landfills after its first use, but components produced using the recycled fibre are themselves recyclable because carbon can retain a significant portion of its virgin properties even after a second reclamation. Furthermore, the recycling process significantly reduces energy costs. Boeing estimates that carbon fibre can be recycled at ≈ 70% of the cost to produce virgin fibre ($8/lb to $12/lb versus $15/lb to $30/lb), using < 5% of the electricity required (1.3–4.5 kWH/lb versus 25–75 kWH/lb). The reason recycling has assumed such a high profile is that airframers, owing to a long history of working with aluminium and other metals, had achieved an enviably high metal-rate of recycling. This fact explains (at least in part) why Boeing and Airbus have been integrally involved in much of the research into carbon fibre recycling over the past several years. Each Boeing 787, for example, carries ≈ 40,000 lb/18,144 kg of salvageable carbon fibre. Boeing, with its industry partners in the Aircraft Fleet Recycling Association (AFRA, Washington, DC, USA) (AFRA, 2006) and Airbus, through its Process for Advanced Management of End-of-Life Aircraft (PAMELA) consortium, are looking to increase the amount of recycled material from aircraft from ≈ 70% today to ≥90% in the coming years.
>
> *(www.compositesworld.com)*

6.1 Methods for recycling

Several recycling methods are in various stages of development (Pimenta and Pinho, 2011; Pickering, 2006; Knight Kouparitsas et al., 2002). Two technology families have been proposed to recycle CFRPs (Fig. 6.1): mechanical recycling and fiber reclamation. Presently, pyrolysis, with and without the use of a catalyst, appears to be the front-running technology (Pimenta and Pinho, 2011; Pickering, 2006). Mechanical recycling reduces the size of the scrap for producing recyclates, whereas thermal process-based

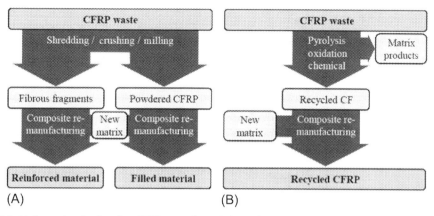

Fig. 6.1 Main technologies for CFRP recycling. (A) Mechanical recycling. (B) Fiber reclamation. *(Reproduced with permission from Pimenta S, Pinho ST. Recycling carbon fibre reinforced polymers for structural applications: technology review and market outlook. Waste Manag 2011;31(2):378–392. https://doi.org/10.1016/j.wasman.2010.09.019.)*

recycling breaks the scrap down into materials and energy. There are currently two continuous production plants in the world, and both are using pyrolysis. One plant is in the United Kingdom and the other in Japan.

6.2 Mechanical recycling

Mechanical recycling also has the advantage of being the most environmentally friendly technology for recycling as no toxic chemicals and gases are used or emitted. Mechanical processing is the first approach considered and adopted for recycling of composite waste. It is also the most commercially viable.

In mechanical recycling the ingredients of the original composite are reduced in size and appear in the resulting recyclates containing mixtures of polymer, fiber, and filler. Usually, this involves a slow-speed cutting or crushing mill for reducing the material to pieces of about 50–100 mm in size. This makes possible the removal of metal inserts, and, if done in an early stage where the waste arises, the volume reduction assists transport. The main size reduction stage is carried out in a hammer mill or in a high-speed mill in which the material is ground into a finer product that has a range from typically 10 mm in size down to particles <50 mm in size. Then a classifying operation (generally using cyclones and sieves) is used for grading the resulting recyclate into fractions of different sizes (Pimenta and Pinho, 2011; Palmer et al., 2009; Scheirs, 1998; Hartt and Carey, 1992; Curcuras et al., 1991).

Normally the finer-graded fractions are powders and contain a higher amount of filler and polymer as compared with the original composite. The coarser fractions are fibrous in nature whereby the particles have higher fiber content and also high aspect ratio. Some companies—ERCOM in Germany and Phoenix Fiberglass in Canada (Scheirs, 1998; Hartt and Carey, 1992; Curcuras et al., 1991)—have been involved in developing recycling activity on a commercial scale (www.tech.plym.ac.uk). These two companies are using the two most common grades of thermoset glass fiber composite material: bulk molding compounds (BMC) and sheet molding compounds (SMC). These composites are based, in general, on polyester resins and contain high amount of filler (example calcium carbonate) or the fire retardant alumina trihydrate.

> ERCOM use a mobile shredder to reduce the initial size. This is expensive equipment and by making it mobile it can be taken to various sites to conduct initial size reduction to increase the bulk density of the material to make transport more cost-effective. The shredder reduces the scrap into small pieces of about 50 × 50 mm in size with a bulk density of ≈ 330 kg/m^3. At a central processing site, a hammer mill is used to break down the scrap material further, and it is graded using cyclones and sieves into several powder and fibrous fractions. A hammer mill is an impact process. It has the advantage that, while there is abrasion on the hammers, there are no blades that require regular sharpening. Phoenix Fiberglass use a similar arrangement that comprises two classifiers.
> **(Pickering, 2006; Sims and Booth, 1993)**

Tables 6.3 and 6.4 show the details of the recyclates produced by ERCOM and Phoenix Fiberglass.

Several applications have been examined for recyclates that in the form of fine powders can be used as substitutes for calcium carbonate filler in new BMC and SMC. Reductions in mechanical properties are tolerable at loading levels of ≈ 10%. But higher proportions give rise to processing problems because recyclates absorb more resin and thereby increase the viscosity of the molding compound, and substantial reductions in mechanical properties are also observed (Scheirs, 1998; Bledzki et al., 1992; Butler, 1991; Jutte and Graham, 1991; Soh et al., 1994). The density of recyclates is lower as compared with calcium carbonate because they contain a substantial amount of low-density polymer. Hence, SMC components containing 10% recyclate as a filler substitute can be 5% lighter as compared with one using only calcium carbonate (Scheirs, 1998). The use of coarser fibrous recyclates in which the larger pieces of recyclate consist of

Table 6.3 Grades of sheet molding compound recyclate from ERCOM GmbH.

Product grade	Fiber length (mm)	Glass content (%)	Bulk density (kg/m^3)
RC1000	0.25	35	670
RC1100	0.25–3	45	460
RC3000	3–15	45	170
RC3101	3–20	45	400

Based on Pickering SJ. Recycling technologies for thermoset composite materials—current status. Compos A: Appl Sci Manuf 2006;37:1206-1215. https://doi.org/10.1016/j.compositesa.2005.05.030.

Table 6.4 Grades of sheet molding compound recyclate from Phoenix Fibreglass, Inc.

Recyclate grade	Particle size (μm)	Glass content (%)	Filler and organic content (%)
PHX-200 filler fraction (PHX-200 is a recyclate grade in a powder form suitable for use as a filler)	14	0.8, 1.6, 3.1	~12
MFX milled fibers (MFX is a recyclate grade in which there is a high glass fraction and the fibers are grades in lengths from 0.8 to 3.1 mm)	13	85	40
CSX hybrid fibers (CSX is a recyclate grade with much larger-sized particles containing proportions of fiber, resin, and filler similar to the original material)	87	15	60

Based on Pickering SJ. Recycling technologies for thermoset composite materials—current status. Compos A: Appl Sci Manuf 2006;37:1206–1215. https://doi.org/10.1016/j.compositesa.2005.05.030; Scheirs J. Polymer recycling, science, technology and applications. Chichester: John Wiley & Sons; 1998.

substantial amounts of intact fiber is found to be more difficult, and reductions in strength properties and toughness have been found even with reasonable additions of fibrous recyclate as a replacement for filler. This phenomenon could be because of the lack of bonding between the recyclate and the polymer and because the larger particles of recyclates act as "stress raisers" in the composite. Treatment of the recyclate for increasing the bonding can improve mechanical properties. In a short-fiber molding compound, fibrous recyclate is used for partially replacing short glass fibers provided the remaining virgin fibers that are replaced with longer fibers. Higher strengths can be achieved with longer virgin fibers, and these can used to counterbalance the harmful effects of the recyclate.

The common applications for mechanically recycled composites include their reincorporation in new composites as filler or reinforcement and use in the construction industry (example as fillers for artificial woods or asphalt or as mineral sources for cement) (Scheirs, 1998; Conroy et al., 2006). But these are the low-value applications. So, mechanical recycling is used essentially for glass fiber-reinforced polymers, although applications to thermoplastic and thermoset CFRP are also reported (ECRC Services Company, 2003; Knight Kouparitsas et al., 2002; Takahashi et al., 2007; Ogi et al., 2007). Individual fibers are not recovered in mechanical recycling, so the mechanical performance of the recyclates is assessed at the composite level.

Environmentally speaking the many steps needed to decrease the size of the fibers to a powder are extremely energy intensive (Morin et al., 2012). However, mechanical recycling does not use any corrosive chemical solvents. Quality wise, this process creates recycled fibers that are extremely short so that their architecture is very unstructured, coarse, and not consistent. As a consequence the mechanical properties are significantly reduced in comparison with VCF (Morin et al., 2012). Furthermore the recyclate is coated in polymer residues, making it difficult to incorporate in new composites and achieve full reinforcement benefits (Pickering, 2009). Overall, mechanical recycling greatly reduces fiber quality and has very limited applications, making it a nonreliable process to implement for wide-scale reuse of carbon composites. It is therefore necessary to evaluate technologies that separate the polymer from the fibers, to preserve length, orientation, and strength.

6.2.1 Thermal/chemical process-based recycling

Thermal/chemical process-based recycling comprises recovering fibers from CFRP using an aggressive thermal or chemical process for breaking down the matrix (generally a thermoset) (Ushikoshi et al., 1995; Gosau et al., 2006; Allred et al., 1996, 1997; Lester et al., 2004; Nahil and Williams, 2011; www.cfk-recycling.com; www.karborek.it/fibra; Cornacchia et al., 2009; www.hadegrecycling.de; Anon, 2010; Pickering et al., 2000, 2009; Yip et al., 2002; Allred et al., 2001; Nakagawa et al., 2009; Piñero-Hernanz et al., 2008; Jiang et al., 2009; Goto, 2009; Marsh, 2009a,b; ATI, 1994; Hitachi Chemical Co. Ltd, 2010; Eckert et al., 1996; Hyde et al., 2006; Warrior et al., 2009; Knight Kouparitsas et al., 2002; Jody et al., 2004; Allen, 2008; Yuyan, 2004; Janney et al., 2009; Li et al., 2012). The fibers are released and collected, and energy or molecules can be recovered from the matrix. This process is especially suitable for CFRP. Carbon fiber possesses high chemical and thermal stability (Pickering, 2006); therefore their excellent mechanical properties (particularly stiffness) are usually not substantially degraded. This process may be preceded by preliminary operations (e.g., cleaning and size reduction of the waste). In general, RCF have a clean surface, and mechanical properties are comparable with their virgin precursors. However, some surface defects and reduction in strength properties have also been observed. After reclamation, recycled fibers are generally reimpregnated with new resin for manufacturing recycled CFRP. Besides, RCF is also being used in nonstructural applications.

Fiber reclamation processes are mainly suitable for CFRPs. Carbon fibers have high thermal and chemical stability (Pickering, 2006); therefore usually their exceptional mechanical properties are not substantially degraded (particularly stiffness). The rCFs usually have a cleaner surface (Fig. 6.2A). The mechanical properties are comparable with the virgin precursors; still, some surface defects (Fig. 6.2B) and strength deterioration (particularly at longer gauge lengths) have also been observed (Heil et al., 2009).

6.2.1.1 Pyrolysis

The thermal decomposition of organic molecules in an inert atmosphere is known as pyrolysis. It is one of the most extensively used recycling processes for CFRP (Pimenta and Pinho, 2011; Pickering, 2006; Ushikoshi et al., 1995; Gosau et al., 2006; Allred et al., 1996; Allred et al., 1997; Lester et al., 2004; Nahil and Williams, 2011;

Recycling of carbon fiber-reinforced polymers   99

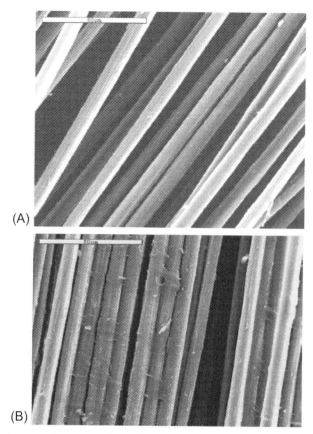

Fig. 6.2 Scanning electron microscopy of recycled (through pyrolysis) carbon fibers. (A) Clean recycled fibers. (B) Recycled fibers with char residue. *(Reproduced with permission from Pimenta S, Pinho ST. Recycling carbon fibre reinforced polymers for structural applications: technology review and market outlook. Waste Manag 2011;31(2):378–392. https://doi.org/10.1016/j.wasman.2010.09.019.)*

www.cfk-recycling.com; www.karborek.it/fibra; Cornacchia et al., 2009; www.hadegrecycling.de; Anon, 2010). During pyrolysis the CFRP is heated up to a temperature of 450–700°C without oxygen. The polymeric matrix gets volatilized into lower-weight molecules, and the carbon fiber remains inert and is finally recovered (Marsh, 2008; Meyer et al., 2007).

Milled carbon has developed a system that recycles both cured and uncured carbon fiber composites using pyrolysis. While a completely oxygen-free process is not possible practically, the intensely oxygen-poor atmosphere maintained during pyrolysis hinders oxidation. This makes possible to recover materials in virtually the same condition as before they were added into the composite. In the process developed by milled carbon, incineration burns off all the resin and additives, freeing the fiber reinforcement. The art of this process is how the atmosphere is controlled. "The process supports the recycling of cured parts up to 2 m/6.56 ft wide, 0.25 m/0.82 ft high and 25 mm/1 inch thick, and uncured material in the form of manufacturing offcuts or unused rolls of pre-impregnated material are processed in a similar fashion. The company does not need to pretreat the material before processing. Milled Carbon can process the recyclate further by shredding, chopping and milling it for various applications. In terms of resin, Milled Carbon typically handle basic epoxies, but the company can also work with epoxies, including bismaleimides and phenolics" (www.compositesworld.com).

Milled carbon is also looking into contaminated EoL waste (e.g., scrapped airplanes), from which any composites come in shredded form with varying degrees of debris accompanying it. The company is working with Nottingham University in the United Kingdom on a fluidized bed process. Exhaust gasses arising from the process could drive new processes, such as the fluidized bed. The composite is immersed in a chemical bath that helps in breaking the material down into its constituent parts. "In the UK, the Recycled Carbon Fibre, Limited (RCF Limited, West Midlands) plant houses a highly sophisticated 120 ft/37 m long pyrolysis machine reportedly capable of recycling $\approx$2,000 metric tonnes (4,409,250 lb) of waste material. The capacity of the system is $\approx$1,200 metric tonnes (2,645,550 lb) of RCF output per year" (Spooner, 2012).

Fig. 6.3 shows RCF derived from continuous pyrolysis. Researchers at Nottingham University have been developing ways of recycling carbon fiber composites for the last decade and have worked with Boeing since 2006 (Pickering, 2006). The final objective is to insert recycled materials back into the manufacturing process, for example, on the plane in nonstructural sustainable interior applications, or in the tooling they use for manufacture. This work helps in creating environmental solutions throughout the life cycle of Boeing products.

Fig. 6.3 Adherent Technologies Inc.'s recycled CF derived from the catalytic conversion process. *(From Personal communication Martin Spooner.)*

The RCF Group is launching a new company called Green Carbon Fibre, which will supply RCF suitable for use in a range of industrial applications. The products Green Carbon Fibre have developed include milled and chopped carbon, as well as carbon fibre pellets for use in thermoplastic applications. All of the products are produced from carbon fibre recovered by its sister company Recycled CF, which will continue to focus on the recycling of carbon fibre waste from the composite industry, taking processing scrap offcuts, moulds and EoL components. The RCF Group operates the world's first commercial-scale continuous carbon fibre recycling plant from its site at Coseley in the West Midlands, UK. The site has the capacity to process 2,000 tonnes of scrap CF composite each year. The RCF Group's second plant will be based in the USA and is expected to be operational soon. Until then, feedstock materials will be collected in the USA and processed at the UK facility.

(www.compositesworld.com)

In Japan, members of the Recycling Committee of the Japan Carbon Fiber Manufacturers Association (JCMA), including Toray Industries Incorporated (Tokyo, Japan), Toho Tenax Company (Tokyo, Japan), a member of the Teijin Group, and Mitsubishi Rayon Company (Osaka, Japan) have formed a joint venture to recycle carbon fibre at a plant owned by Mitsui Mining Company (Omuta City, Japan). They started working on CFRP recycling in 2006. A test plant began operation in 2007, followed by verification in 2008. Reportedly, the annual output of RCF at the plant will ramp-up from several hundred metric tonnes initially to 1,000 metric tonnes (2.2 million lb) as demand increases. The recyclate, after compounding, is targeted primarily at the consumer electronics and automotive industries.

(www.compositesworld.com)

Ushikoshi et al. (1995) applied pyrolysis for the recycling of carbon fiber. Carbon fiber sheets were heated in a stream of air at 400°C, 500°C, and 600°C. Initially the fibers are protected by the resin, so they do not oxidize. Oxidation would be detrimental for the fibers because it reduces their strength.

Ushikoshi et al. (1995) stated that carbon fiber begins to oxidize only when almost all the resin has disappeared. Then the process is stopped, and the recovered fibers remain at almost 100% of their original strength. This process has been described as "unusual" by Pickering (2006) and Meyer et al. (2007). The results suggested that there was the least degradation in tensile strength at 500°C (Table 6.5).

In the United States, Adherent and its development partner, Titan Technologies (Albuquerque, New Mexico), have developed a catalytic pyrolysis process (ATI, 1994; Gosau et al., 2006; Allred et al., 1996, 997) for recycling complex mixtures of thermoplastic and cross-linked thermoset polymers and also automobile/truck tires and mixed electronic scrap. Pyrolization takes place at a lower temperature (~200°C) in the presence of a proprietary catalyst. The polymer is completely degraded into low-molecular weight hydrocarbons in liquid or gaseous form, and the remaining carbon fiber are substantially free from resin.

Fig. 6.4 shows the Phoenix reactor of ATI, a pilot-scale vacuum pyrolysis unit. Fig. 6.5 shows a low-temperature, low-pressure recycling reactor. Fig. 6.3 shows adherent RCF derived from the catalytic conversion process at Technologies Incorporated.

Table 6.5 Tensile properties of carbon fiber.

	Original carbon fibers T300	Original carbon fibers M40	T300/Epoxy pyrolysis at 500°C	T300/Epoxy pyrolysis at 500°C	Orig. T300 fibers heated in air at 500°C	Orig. M40 fibers heated in air at 500°C
Tensile strength (MPa) (and degradation)	3660	3360	3370 (−7.9%)	3130 (−6.8%)	~2000 (−45%)	~2500 (−26%)
Tensile modulus (GPa) (and degradation)	241	395	194 (19.5%)	355 (−10.1%)	–	–

Based on Ushikoshi K, Komatsu N, Sugino M. Recycling of CFRP by pyrolysis method. J Soc Mater Sci 1995;44 (499):428–431.

Fig. 6.4 ATI's Phoenix reactor, a pilot scale vacuum pyrolysis unit. *(Reproduced with permission from Gosau JM, Wesley TF, Allred RE. Proceedings of the 38th SAMPE technical conference, Dallas, TX, United States, 7–9th November, 2006; 2006.)*

This 13-year effort has benefited from support from the US Government—about $3 million (USD) for R&D but the partners are now finding investors who can help moving the company from its prototype work into a commercial-scale offering. Adherent contends that a main cost barrier in composite recycling is that collected composite waste must be sorted. This is one of the more labor-intensive aspects of traditional recycling processes.

Fig. 6.5 Low-temperature, low-pressure recycling reactor. *(Reproduced with permission from Gosau JM, Wesley TF, Allred RE. Proceedings of the 38th SAMPE technical conference, Dallas, TX, United States, 7–9th November, 2006; 2006.)*

Its reclamation process is designed for treating all materials at once in a tertiary process involving thermal pretreatment and two wet chemical processes. The term tertiary recycling is coined by the American Plastics Council. This process recovers the fibers and also thermoplastic and thermoset polymeric waste, the latter in the form of reusable hydrocarbon fractions that may be used later for producing new polymers, monomers, fuels, or other chemicals.

In this process, there is no need to sort and separate the materials, thereby removing transportation, labor, and other costs. All scrap parts and other materials are introduced to the recycling process as a single feedstock. The materials are separated by designated support unit operations during the process before and after entering the tertiary recycling reactor. The support units handle size reduction using crushing and chopping, drying, material classification, off-gas treatment, and distillation, as well as the recovery of metals, fiber, and carbon char. Although this process can recycle only a few hundred pounds of scrap a day, the purity of recovered carbon fiber is 99.9%. Adherent has performed single-fiber tests for comparing virgin carbon fiber fabric with recovered fibers. Reclaimed fiber showed reduction in tensile strength of only 8.6%, supporting Adherent's argument that reclaimed fibers significantly retain their properties (https://www.compositesworld.com/articles/carbon-fiber-life-beyond-the-landfill).

> *Once the company ramps up to full commercial capacity, it expects to erect a plant that will be able to process 4,000 lb/1,814 kg a day. The company has forecast that it has the potential to recover ≈435,000 lb/162,359 kg of high-purity carbon fibre from 66,000 lb/246,338 kg of carbon composite scrap in its first year. At this volume, the plant would be running at ≈40% capacity, permitting significant growth in subsequent years.*
>
> *(www.compositesworld.com)*

Microwave pyrolysis is another form of CFRP recycling under development by companies and universities in the United States, United Kingdom, and Germany (Allen, 2008). Microwave heating is the process of coupling materials with microwaves, absorbing electromagnetic energy and transforming it into heat within the material volume, in which the heat is generated from the inside to the entire volume (Khaled et al., 2018; Oghbaei and Mirzaee, 2010; Yadoji et al., 2003). The microwave method can treat uniform samples due to its features of volumetric and internal heating (Rybakov and Buyanova, 2018; Kumar et al., 2018). Fig. 6.6 shows schematic of carbon fiber-reinforced polymer (CFRP) waste treatment by microwave thermolysis.

Fig. 6.7 shows the recovered carbon fibers after traditional thermolysis (Fig. 6.7A–C) and microwave thermolysis (Fig. 6.7D).

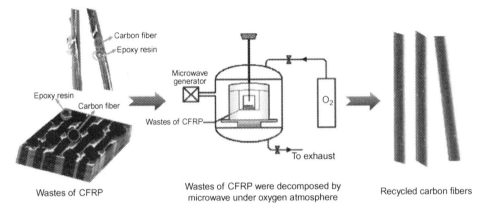

Fig. 6.6 Schematic of carbon fiber-reinforced polymer (CFRP) waste treatment by microwave thermolysis. *(Reproduced with permission from Deng J, Xu L, Zhang L, Peng J, Guo S, Liu J, Koppala S. Recycling of carbon fibers from CFRP waste by microwave thermolysis. Processes 2019;7:207.)*

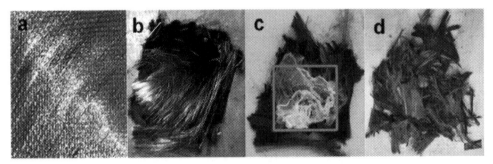

Fig. 6.7 The macroscopic appearances of the CFRP waste samples after experiments (A) 30 min of traditional heating at 400°C, (B) 30 min of traditional heating at 450°C, (C) 30 min of traditional heating at 500°C, and (D) 13 min of microwave heating at 450°C. *(Reproduced with permission from Deng J, Xu L, Zhang L, Peng J, Guo S, Liu J, Koppala S. Recycling of carbon fibers from CFRP waste by microwave thermolysis. Processes 2019;7:207.)*

Long et al. (2015) observed that microwave irradiation is a flexible, easily controlled, efficient method to recover high-value carbon fibers, and recovered carbon fibers could be directly used as reinforcement in new polymers (polypropylene and nylon). This shows the great potential of effectively recycling carbon fiber by the microwave technique.

Lester et al. (2004) experimented on the application of microwaves for heating the composite waste to break down the matrix into gases and oil in an inert atmosphere. Heating with microwaves is found to be advantageous because the waste composites get heated within its core at a fast thermal transfer rate and therefore ensures some energy savings.

Deng et al. (2019) successfully recovered carbon fibers by thermolysis under an oxygen atmosphere. Use of microwave thermolysis for recovering carbon fibers from CFRP waste is an attractive approach. In comparison with the conventional method, there was a reduction in reaction time by 56.67% and increase in recovery ratio by 15%. Microwave thermolysis is fast and more efficient, needs less energy, and obtains cleaner recovered carbon fibers as compared with those recovered using the conventional thermolysis process.

In general, microwave energy absorbed by the conductive properties of carbon fibre heat the matrix resin internally rather than externally. This can result in more rapid decomposition of resin and recovery of fibres without char formation, shorter overall processing time, and smaller-scale equipment than is required for other pyrolysis methods. Over the past 3 years, the research company Firebird Advanced Materials Incorporated (Raleigh, NC, USA) have built a small pilot-scale installation to test their microwave recycling process and, this year, have begun implementation of their commercialisation plan.

(Lester et al., 2004; McConnell, 2010)

Composite waste containing carbon fibre and polybenzoxazines resin was pyrolysed in a fixed bed reactor at temperatures of 350 to 700 °C. Solid residues of between 70 wt% and 83.6 wt%, liquid yields 14 wt% and 24.6 wt% and gas yields 0.7 wt% and 3.8 wt% were obtained depending upon the pyrolysis temperature. The pyrolysis liquids obtained contained aniline at high concentration along with oxygenated and nitrogenated aromatic compounds. The pyrolysis gases contained mainly of carbon dioxide, carbon monooxide, methane, hydrogen and other hydrocarbons. The carbon fibre used in the composite waste were separated from the char of the solid residue through oxidation of the char at two temperatures and their mechanical strength properties were determined. The carbon fibre recovered from the sample pyrolysed at 500 °C and oxidised at 500 °C showed mechanical properties which were 90% in comparison with those of the original virgin carbon fibre. Steam activation of the recovered carbon fibre was conducted at 850 °C at different times of activation. A maximum BET surface area of $>800\ m^2\ g^{-1}$ was achieved for the activated carbon fibre produced at 850 °C after 5 hours of activation. Nitrogen adsorption–desorption isotherms revealed that the adsorption capacity increased as the activation time increased up to 5 hours of activation, and then reduced after that.

(Nahil and Williams, 2011)

In the United States, Materials Innovation Technologies RCF (MIT-RCF) started recycling CFRP in 2008 using an undisclosed pyrolysis process. The first step involved chopping the feedstock to a consistent length after pyrolysis. "An in-house manufacturing process proved to be particularly suitable for re-manufacturing. In Germany, CFK Valley Stade Recycling GmbH, and Company, KG (2007) (www.cfk-recycling.com) use a continuous pyrolysis process. This process has been developed with the Technical University of Hamburg-Harburg and ReFiber ApS. The process is suitable for several types of CFRP waste. The main products comprise milled fibres, chopped fibres, and textile products. This process is complemented with an oxidation step for char removal" (www.cfk-recycling.com).

In Italy, Karborek (1999) uses a combined pyrolysis and upgraded (in oxygen) patented process for recycling the fibers and avoiding char formation (Cornacchia et al., 2009). Fiber length is maintained during reclamation. The major products of Karborek's are milled and chopped RCF and also blended nonwoven veils with carbon and thermoplastic fibers.

In Germany, Hadeg Recycling Limited (HADEG Recycling GmbH, 1995), worked jointly with the Technical University of Hamburg. This company reclaimed carbon fiber by pyrolysis, and also commercialized unprocessed manufacturing remains (dry carbon fiber rovings or fabrics and uncured prepreg cutoffs).

ELG Haniel of Germany processing scrap metals has acquired 100% of the shares in Recycled Carbon Fiber, a UK company. CFRP production waste is converted into reusable carbon fiber. ELG has plans of expanding the business worldwide. RCF has set up its UK processing facility, which can process 2000 tonnes/year (t/y) of carbon fiber waste. In 2011 the company signed a contract with GKN Aerospace under which will recycle the uncured carbon fiber composite waste from its aerostructure manufacturing operation in Cowes, Isle of Wight. Table 6.6 and Fig. 6.8 show the product range of ELG Carbon Fiber Limited., and Fig. 6.9 shows the recycling furnace (Spooner, 2012).

ELG has taken out European and the US patents (EP 2152487 and US Patent 12/669425). In Europe, Airbus is leading the PAMELA sustainability project of European Union. In southwest France a test facility has been established, where Airbus intends to prove that more than 85% of every aircraft can be recycled, reused, or otherwise recovered. Boeing is heavily involved in carbon fiber reclamation research and helped form the Aircraft Fleet Recycling Association (AFRA), a consortium of the US and European

Table 6.6 Product range of *ELG Carbon Fiber Ltd.*

Milled fibers—adjustable average filament lengths of 100–4001
Chopped fibers—random fiber lengths 6–60 mm
Milled fiber pellets—with binder
Chopped fiber pellets—with binder

From Personal communication, Martin Spooner.

Fig. 6.8 Product range of *ELG Carbon Fibre Ltd. (Reproduced with permission from Personal communication Martin Spooner.)*

Fig. 6.9 Recycling furnace (ELG carbon Fibre Ltd.). *(Reproduced with permission from Personal communication, Martin Spooner.)*

companies. AFRA is focusing on recovery of carbon fiber from manufacturing waste and retired airplane scrap. Boeing is developing a commercial carbon fiber recycling reactor in the United States. Milled carbon and Adherent Technologies are the partners (PAMELA, 2008).

Trek Bicycle has started a full-scale carbon recycling program at their Waterloo-based manufacturing facility in the United States and is now recycling all scrap carbon fiber. Through a partnership with Materials Innovation Technologies (MIT, Fletcher, North Carolina, United States) and their wholly owned subsidiary MIT-RCF (a South Carolina carbon fiber reclamation facility), Trek completed a 3-month trial period and determined the viability of using recycling as an official part of the manufacturing process. Trek collects excess trimmings and noncompliant molded parts and combines it with select reclaimed warranty frames to send to MIT's South Carolina facility to start reclamation. Reclaimed carbon fiber is used in reinforced thermoplastic applications. R&D is going on for use in automotive, aerospace, medical, and recreational applications. Carbon fiber recycling holds enormous potential for the whole industry. Boeing plans to invest USD $1 million per year in a strategic research collaboration with Nottingham University to improve the recycling of aircraft components made from carbon fiber-reinforced plastics. The aircraft manufacturer has agreed to fund the work. This support will allow Nottingham University to carry out more technology development, with the aim of processing recycled fiber in new applications and products in collaboration with suppliers. The president of Boeing UK commented that the final objective is to insert recycled materials back into the manufacturing process, for example, on the plane in nonstructural sustainable interiors applications, or in the tooling they use for manufacture. This work is helping to create environmental solutions throughout the life cycle of Boeing products. Based on their rich knowledge of material recycling, Takayasu KK has developed carbon fiber recycling technology (Anon, 2010). The technology enables the recovery of carbon fiber from CFRP and prepregs by optimized thermal decomposition. The resulting RCF (Takayasu Recycle Carbon Fiber [TRCF]) is in a staple format and retains 90% of its original strength and elastic modulus. TRCF is also electrically conductive and offers a wide range of end-use applications as nonwovens and sheets. Takayasu KK also provides polyester staple containing an organic antibacterial agent mixed into the resin. Nonwovens made from this staple display antibacterial activities against a wide spectrum of bacteria, including *Trichophyton*, *Escherichia*, and *Staphylococcus*.

6.2.1.2 Oxidation in fluidized beds

The fluidized bed process (FBP) has been developed and implemented by Pickering et al. (2000) at the University of Nottingham (Fig. 6.10).

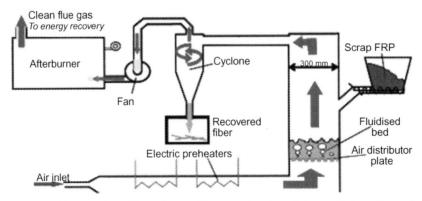

Fig. 6.10 Fluidized bed method. *(Reproduced with permission from Pickering SJ, Recycling technologies for thermoset composite materials—current status. Compos A: Appl Sci Manuf 2006;37:1206–1215. https://doi.org/10.1016/j.compositesa.2005.05.030.)*

This process has been developed for the recovery of glass fiber and carbon fiber. The advantages of this process are the following:
- high tolerance to contamination,
- no residual char on the fiber surface,
- well-established and documented process.

The drawbacks are the following:
- strength degradation between 25% and 50%,
- fiber length degradation,
- unstructured ("fluffy") fiber architecture,
- the impossibility of material recovery from resin.

> *During recycling, CFRP scrap reduced to fragments of about 25 mm in size is fed into a bed of silica with a particle size of about 0.85 mm. The sand is fluidised with a stream of hot air, and typical fluidising velocities are 0.4–1.0 m/s at 450–550 °C. In the fluidised bed, the polymer volatilises from the composite. This action releases the fibres and fillers to be carried out of the bed as individual particles suspended in the gas steam. The fibres and fillers are then separated from the gas stream, which can then pass into a high-temperature secondary combustion chamber where the polymer is fully oxidised. Energy may subsequently be recovered from these hot combustion products. The fibre product is in a fluffy form comprising individual fibre filaments of mean length (by weight) from 6 mm to > 10 mm. The fibres are clean and show very little surface contamination. A glass-reinforced polyester composite can be processed at 450 °C, at which temperature the polymer volatilises and releases the fibres into the gas stream. Epoxy resins require ≤550 °C for rapid volatilisation of the polymer. Glass fibres typically suffer a 50% reduction in tensile strength but retain the same stiffness as the virgin fibre if processed at 450 °C. At higher temperatures, there is significantly greater reduction in mechanical strength, resulting in a 90% reduction in strength at 650 °C. Carbon fibre show a lower strength degradation of typically 20% with retention of the original stiffness if processed at 550 °C. Even though it is processed in air, the carbon fibre do not show measurable oxidation. Analyses of the surface of the RCF also shows that there is only a small reduction in surface oxygen content, indicating that the fibres have good potential for bonding to a polymer matrix if re-used in a composite. A particular advantage of the FBP is that it is very tolerant of mixed and contaminated materials.*
>
> **(Pickering, 2006)**

Yip et al. (2002) developed a carbon fiber recycling process for scrap composites on the basis of fluidized bed technology. The recycling process is described, together with the characterization methods used to analyze the quality of recycled fiber. Recycled fibers of a mean length of ≤10 mm were recovered, and they retained about 75% of their tensile strength. The Young's modulus did not change, and the surface condition was similar to that of the virgin fiber.

The fluidized bed process (Table 6.7), however, produces carbon fiber with almost no reduction in modulus and 18%–50% reduction in tensile strength as compared with virgin carbon fiber making it suitable for applications needing high stiffness rather than strength.

Table 6.7 Fluidized bed method.

- *Advantages*
  - Clean high-quality fibers
  - Good mechanical property retention
  - Process tolerant of contamination
  - Process simple and easily scalable
- *Disadvantages*
  - No material recovery from polymer (33% material and 66% energy)
  - Some strength degradation (~25%)

Based on Turner T, Warrior N, Pickering S. High value composite materials from recycled carbon fibre. www.nottingham.ac.uk/HIRECAR; 2010.

6.2.1.3 Chemical recycling

Chemical methods are based on a reactive medium. Catalytic solutions, benzyl alcohol, and supercritical fluids (SCF) are used at low temperature (usually <350°C) (Allred et al., 2001; Nakagawa et al., 2009; Piñero-Hernanz et al., 2008; Jiang et al., 2007, 2008, 2009; Goto, 2009; Marsh, 2009a,b; ATI, 1994; Zhu et al., 2019; Li et al., 2012; Wang et al., 2015; Wong et al., 2009; Loppinet-Serani et al., 2010; Maxime, 2014; Xu et al., 2013; Witik et al., 2013; Morin et al., 2012; Pimenta and Pinho, 2011; Anane-Fenin and Akinlabi, 2017; Bai et al., 2010; Feraboli et al., 2012; Job, 2014; Knight, 2013; La Rosa et al., 2016; Markets and markets, 2017; Okajima et al., 2012; Okajima and Sako, 2017; Oliveux et al., 2015; Onwudili et al., 2013; Yan et al., 2016; Yuyan et al., 2009). The polymeric resin is decomposed into relatively large and so high-value oligomers, whereas the carbon fiber remains inert and is collected afterward. The mechanical properties and fiber lengths are very well retained, and there is a high potential for material recovery from resin. The drawbacks are low tolerance to contamination, reduced adhesion to polymeric resins, and possible impact on the environment if hazardous solvents are used.

Researchers in Japan developed a carbon fiber-epoxy recycling process using benzyl alcohol and a catalyst in an atmosphere of nitrogen. The plant includes a distillation system for cleaning the reaction fluid (which is then reintroduced in the system) and recovering resin-based products. This process has reclaimed carbon fiber from EoL components from the aeronautics and sports industries (Pimenta and Pinho, 2011; Hitachi Chemical Co. Ltd, 2010).

SCF are fluids at temperatures and pressures generally just above the critical point. These fluids can diffuse through solids like a gas and dissolve materials such as liquids and are suitable as substitutes for organic chemical solvents in various processes (Fig. 6.11).

"The fluid presents itself in a single supercritical phase at this stage while possessing combined characteristics: liquid-like density and dissolving power, and gas-like viscosity and diffusivity. So SCF can penetrate porous solids and dissolve organic materials, while

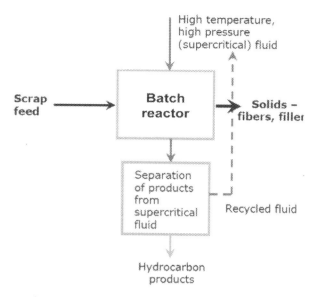

Fig. 6.11 Supercritical fluid method. *(Reproduced with permission from Pickering SJ. Recycling technologies for thermoset composite materials—current status. Compos A: Appl Sci Manuf 2006;37:1206–1215. https://doi.org/10.1016/j.compositesa.2005.05.030.)*

Table 6.8 Supercritical fluid method.

- *Advantages*
- – Clean high-quality fibers
- – Excellent mechanical property retention
- – Potential for long fiber recovery
- – Material recovered from polymer
- *Disadvantages*
- – Process complexity
- – Still under development

Based on Turner T, Warrior N, Pickering S. High value composite materials from recycled carbon fibre. www.nottingham.ac.uk/HIRECAR; 2010.

being relatively harmless under atmospheric conditions" (Eckert et al., 1996; Hyde et al., 2006). Various types of SCF are being used for carbon fiber recycling, like water, ethanol, methanol, propanol, and acetone (Hyde et al., 2006; Piñero-Hernanz et al., 2008).

These have usually been coupled with alkali catalysts. Chemical recycling with SCF is a current approach. Nevertheless, it is known for producing RCF with almost no mechanical degradation, particularly if using propanol and enabling the recovery of useful chemicals from the matrix (Pickering, 2009; Warrior et al., 2009).

Table 6.8 shows the advantages and disadvantages of supercritical fluid method.

The effect of alkali and alcohol for the decomposition of CFRP in supercritical water was reported by Sugeta et al. (2001). Resol resin as a matrix resin in CFRP was used. Five percent of resol resin was decomposed without using any catalyst at 380°C and 30 min. By the addition of 2 M sodium hydroxide, the decomposition ratio increased to 66.8%. Furthermore the combination of 70-vol% ethanol and 2-M potassium hydroxide accelerated the decomposition of the matrix resin by up to 93.3%.

The use of sulfuric acid as a catalyst in subcritical water was studied by Yuyan et al. (2009). The matrix resin of CFRP was bisphenol-A-type epoxy resin, which was cured with isophorone diamine. The decomposition ratio of the epoxy resin was 58.9% at 260°C and 75 min of reaction time without using any catalyst. Contrarily, it increased to 91.5% in 75 min using 1 M sulfuric acid solution as a catalyst and decomposed completely in 90 min. The tensile strength of the recovered carbon fiber reduced to 1.8%, in comparison with that of virgin one.

The recycling of CFRP in supercritical water using oxygen was studied by Bai et al. (2010) at $440 \pm 10°C$ and 30 ± 1 MPa. The matrix resin was epoxy resin that was cured with phthalic anhydride and benzyl dimethylamine. The recovered carbon fiber showed a higher tensile strength as compared with the virgin fiber when the decomposition ratio was above 85% at 30 ± 5 min of reaction time. However, the tensile strength of the recovered carbon fiber reduced speedily when the decomposition ratio increased beyond 96.5% because of the excessive oxidation of the recovered carbon fiber.

Liu et al. (2012) studied the combined effect of phenol and potassium hydroxide as a catalyst in subcritical water. The matrix resin was bisphenol-A-type epoxy resin hardened with diaminodiphenylmethane. The optimum ratio of potassium hydroxide weight/phenol weight/water volume was 1 g/10 g/100 mL, and the decomposition efficiency of epoxy resin reached about 100% at 325°C, 10.5 MPa, and 20 min. There was no appreciable drop in tensile strength of the recovered carbon fiber in comparison with that of virgin one.

The feasibility of recycling of CFRP using supercritical water was investigated by Princaud et al. (2014). The decomposition of matrix resin in CFRP by hydrolysis and recovery of carbon fiber was evaluated from the view point of the environmental evaluation and the economic validation, based on life-cycle assessment.

> *Researchers at the University of Nottingham have raised hopes that a new system they have shown at small scale will facilitate the recovery, in near-virgin condition, of long high-modulus fibres that can be reused in significant structures. The answer is to chemically dissolve the epoxy matrix away. The researchers were able to identify a solvent which is manageable and effective, leaving the stripped fibres with almost all the strength and stiffness they possessed originally. Its method can also recover useful chemicals from the epoxy resin.*
>
> ***(Marsh, 2009a,b; www.reinforcedplastics.com)***

Knight et al. (2010) have used supercritical water with 0.05 M potassium hydroxide as the catalyst for recycling of an aerospace-grade high-performance epoxy carbon fiber composite (Hexcel 8552/IM7). In this process, nearly 99.2 wt% resin elimination was obtained. This resulted in the recovery of clean, undamaged fibers. The recovered fibers retained the original tensile strength. The feasibility of recycling multiple layer composites was also studied. The chemical recycling processing is the route with the most space for innovation and achieving an environmentally friendly, sustainable, energy saving, and financially rewarding process. The most significant advantage of this process to both mechanical and thermal processes is the quality of recyclate fiber recovered. The mechanical properties of recyclate fibers are close to or in some case the same as virgin fibers. This means that they can be used in the manufacture of new thermoset composites while avoiding the cost associated with the production of virgin fibers. The drawbacks to this process include the use of chemicals that are toxic and the high cost of reactor (Anane-Fenin and Akinlabi, 2017).

The recycling of a CFRP composite by using an aqueous electrolyte solution in the presence of electrical currents was studied. "The C-N bonds of the epoxy resin can be broken in an electrically driven heterogeneous catalytic reaction, leading to the decomposition of the resin component and recovery of CFs via the benefit of the EPOC effect. The properties of the rCFs are dependent on the combined effect of all the variables. By optimizing the working conditions, depolymerization of epoxy resin and reclamation of carbon fibres are found to be efficient even under ambient temperature and pressure using conventional and nontoxic chemicals. This method achieves a high (nearly 100%) extent of epoxy resin removal. The average residual tensile strength and IFSS are approximately 90% and 120%, respectively, of those of VCFs. Compared with existing recycling methods, this new method is simple and green" (Zhu et al., 2019) (Table 6.9).

"The commercial value of the recycled fibres based on the proposed technology is expected to dramatically increase, since (1) the simplicity of the procedure and low facility demand will remove the size restrain on the composite waste; and (2) the dimensions and strength of CFs can be maintained in the recycling procedure. In addition, this technology can be implemented in an overall cost-effective manner on a large scale and will profoundly impact the management of CFRP composites and possibly other thermoset waste materials" (Zhu et al., 2019). The researchers showed that even using simple equipment and traditional nontoxic electrolyte components intact carbon fibers can be recycled under atmospheric pressure and room temperature in an efficient manner. Especially the simplicity of the process and the moderate needs of the processing facilities remove the CFRP waste size limit, greatly improving the commercial value of recycled fibers and allowing implementation on a large scale (pubs.rsc.org).

Table 6.9 Comparison between existing recycling methods and the new EHD method.

Assessment		Methods						
		Mechanical Milling	Thermal Combustion	Pyrolysis	Fluidized bed	Chemical Solvolysis	Supercritical/subcritical	EHD method
Facility	Temperature	Room temperature	1400–1600°C	400–1000°C	450–500°C	~90°C	400–610°C	Room temperature
	Pressure	Atmospheric pressure	Atmospheric pressure	Atmospheric pressure	10–25 kPa	Atmospheric pressure	22.1–35 MPa	Atmospheric pressure
	Toxicity	Nontoxic	Nontoxic	Nontoxic	Nontoxic	Toxic	Toxic	Nontoxic
	Waste size limit (length)	10–50 mm	NA	6–25 mm	5–10 mm	1 μm–50 mm	10–50 mm	No limit
Performance	Residual tensile strength (vs VCF)	50%–65%	NA	50%–85%	10%–75%	85%–98%	85%–98%	89.46%
	Shear strength (vs VCF)	Unreported	NA	NA	80%	Unreported	Unreported	120.74%
Environmental issues		Dust	Pollutant gas, dust, high energy use	Pollutant gas, high energy use	Pollutant gas, organic solvent, high energy use	Organic solvent	Organic solvent, high energy use	None

Based on Zhu JH, Chen Pi-yu, Su M, Peia C, Xing F. Recycling of carbon fibre reinforced plastics by electrically driven heterogeneous catalytic degradation of epoxy resin Green Chem 2019;21:1635–1647.

6.2.1.4 Other methods

Other methods such as chemical treatment, molten salt treatment, and thermal shock treatment have also been tried. The chemical degradation treatment exposes CFRP to solvents, acids, and bases at various temperatures to degrade and break down the polymeric substrate to liberate the fibers (Jody et al., 2004; Allen, 2008).

Study of this process revealed that different chemicals were required to break down the different polymers, rendering the process economically infeasible. The fibers are rinsed with water after the process is completed, and chemical waste is also a problem. Another technical article investigated the use of reclaiming carbon fiber using a solvent method in nitric acid solutions. The study optimized the temperature of the solution (90°C) with a nitric acid concentration of 8 M and a ratio of the sample weight to the nitric acid solution volume of 4 g:100 mL (Yuyan, 2004). Electron probe microscopy found the RCF to be clean and undamaged. Reclaiming carbon fiber by molten salt processing at 400–600°C uses molten or dissolved salt baths to separate carbon fiber from the polymer matrix (Spooner, 2012). This research is part of the Recycomp project evaluating composite waste thermoset matrices. Scanning electron micrographs of carbon fiber recovered from thermoset matrices using molten salt treatment show that a large amount of resin/residue and particulates remain on the surface of the fibers. Thermal shock treatment attempts to recover carbon fiber from polymer matrices (Jody et al., 2004). The composite panels are dipped into liquid nitrogen and then placed immediately into boiling water. The process, in theory, attempts to leverage the differences in the coefficients of thermal expansion between the polymer and carbon fiber. However, this method usually fails to produce clean fibers because visible separation between the two constituent materials is not observed.

Use of carbon fiber-reinforced composites continues to increase in industries such as the aerospace and automobile industries, where weight reductions lead to reduction of fuel consumption and emissions. For the world to continue benefitting from this material, increasingly more strict environmental regulation must be met, thus requiring proper disposal of carbon fiber at end of life.

Current and most optimal technology for meeting commercial and environmental constraints is a thermal process called pyrolysis. The process of pyrolysis surpasses any other recycling methods in terms of ecological and economical impact. Pyrolysis yields slightly more toxic waste than mechanical methods yet is less energy intensive than both alternatives. This thermal process also produces much greater fiber quality than mechanical recycling and similar quality to chemical recycling. Unlike chemical recycling, pyrolysis can retrieve great quality RCF for a wide range of CFRP waste products. Industries worldwide seeking to recycle CFRPs should therefore use pyrolysis processes over chemical or mechanical methods, since it is clearly the best solution available today.

The mechanical properties of recycled carbon fibre are generally good and compare well with virgin fibre. Measurements of stiffness show that the recycling processes generally gives recycled fibres with a stiffness similar to that of virgin fibre. However, there is some degradation in tensile strength. The thermal fluid processes only show a reduction of a few percent. The pyrolysis processes are reported to show slightly more strength reduction with up to 10% loss in strength. However, the fluidized bed process shows a loss of strength of up to 50%. This may be due to the increased mechanical agitation in the process and also the effect of the oxidizing atmosphere. Though process is particularly suitable for contaminated end-of-life waste which may not be appropriate for other recycling processes. The electrical conductivity of the recycled fibre has been found to be similar to that of virgin fibre and analysis of the surface chemistry shows that after recycling there are still active oxygenated species on the surface and the recycled fibres have been found to bond well to epoxy resin.

(Pickering et al., 2016)

6.3 Energy consumption of the recycling processes

Today energy conservation has become a top priority for business organizations for increasing profitability and deriving genuine strategy over competitor to win more new customers. Apart from the capital cost, the amount of energy consumed for recycling a kilogram of carbon fiber has been an important factor in selecting a profitable recycling process. For the carbon fiber recyclers, this information is confidential (Wong et al., 2017). Nonetheless a few suggestions can still be found, which were mostly obtained from modeling with conservative assumptions. Table 6.10 presents the estimated energy spent in recovering 1 kg of carbon fiber using different recycling techniques. Substantial

Table 6.10 Estimated energy consumption of the main recycling processes.

Mechanical size reduction	0.3 (shredding rate: 150 kg h^{-1})
	2.0 (shredding rate: 10 kg h^{-1})
Fluidized bed process	6 (feed rate: 12 kg h^{-1} m^{-2})
	40 (feed rate: 3 kg h^{-1} m^{-2})
Pyrolysis	2.8 (feed rate unavailable)
	30.0 (feed rate unavailable)
Solvolysis	63 (17,000 tennis racket/month)
	91 (1000 tennis racket/month)

Based on Wong K, Rudd C, Pickering S, Liu X. Composites recycling solutions for the aviation industry citation. Sci China Technol Sci 2017;60:1291. https://doi.org/10.1007/s11431-016-9028-7; Shibata K, Nakagawa M. CFRP recycling technology using depolymerization under ordinary pressure. Hitachi chemical technical report; 2014. p. 6–11; Howarth J, Mareddy SSR, Mativenga PT. Energy intensity and environmental analysis of mechanical recycling of carbon fibre composite. J Clean Prod 2014;81:46–50; Meng F, Mckechnie J, Turner TA. Energy and environmental assessment and reuse of fluidised bed recycled carbon fibres. Compos A: Appl Sci Manuf 2017;100:206–214; Song YS, Youn JR, Gutowski TG. Life cycle energy analysis of fiberreinforced composites. Compos A: Appl Sci Manuf 2009;40:1257–1265; Witik RA, Teuscher R, Michaud V, Ludwig C, Manson JAE. Carbon fibre reinforced composite waste: an environmental assessment of recycling, energy recovery and landfilling. Compos A: Appl Sci Manuf 2013;49:89–99. https://doi.org/10.1016/j.compositesa.2013.02.009.

energy saving is found from each cases as compared with the production of virgin polyacrylonitrile-based carbon fiber, which needs 245 MJ kg^{-1} for producing polyacrylonitrile precursor and then another 459 MJ kg^{-1} for conversion into carbon fiber (Das, 2011). The numbers in the table are only indicative and might have significant deviation from the real production process as constant improvement has always been practiced in industry for cost saving.

ELG RCF Ltd. had successfully reduced energy consumption by 35%. Nevertheless, pyrolysis has been found to be the most extensively used process. Several pyrolysis plants have been reported (Wong et al., 2017):

- Vartega Inc., United States;
- Carbon Conversion Inc., United States;
- ELG Carbon Fibre Ltd., Cosley, United Kingdom;
- CFK Valley Stade Recycling GmbH & Co KG, Germany.

In response to evergrowing demand of carbon fiber, several commercial organizations have setup an intense carbon fiber recycling programs of their own, mainly in North America and European Union (Gardiner, 2014; Pervaiz et al., 2016). This subject has become an important strategic issue that a bill was introduced in the US Senate requesting a study of the technology and energy savings of recycled carbon fiber. The bill, S. 1432, the Carbon Fiber Recycling Act of 2015, directs the DOE to work with the automotive and aviation industry for developing a recycled carbon fiber demonstration project (Caliendo, 2015).

Some of the important initiatives are presented in the succeeding text:

- "SGL-Germany: A BMW-SGL joint venture, SGL Automotive Carbon Fibers, has developed a recycling process to recirculate carbon fibers into the production process. A new material class *Recycled carbon fibers RECAFIL* has been introduced in the form of a Carbon Fiber Cut Mix or as so-called Carbon Fiber Flocks. SGL recycles CF scrap from weaving into non-woven mats and subsequently molds these materials into rear seat and roof structures of BMW i3 models, (Gardiner, 2014; https://www.press.bmwgroup.com/global/startpage.html; https://www.sglgroup.com/cms/international/products/product-groups/cf/recafil/index.html?__locale=en).
- CFK Valley Recycling (Stade, Germany) is another major player in reclaiming carbon fibre, particularly from aerospace industry. The fiber reclaimed by current means is chopped and not suitable for use in wind turbine and aircraft structures. However, discontinuous fiber has long been a favorite of automotive composites, especially interiors and under the hood applications.
- MIT-LLC, USA, started reclaiming carbon fibres 2009 from different industrial waste streams through its own indigenous processes. The reclaimed carbon fibres are transformed into wet-laid nonwoven preforms measuring in widths up to 49 inch and weighing 50 to 1000 g/m^2. MIT uses its Three Dimensional Engineered Preform (3-DEP®) chopped fiber composite technology, developed under DOE-USA Small Business Innovation Research (SBIR) project, to address the need for cost-effective, high volume, lighter weight components for automobiles" (file.scirp.org).

6.4 Energy and environmental benefits at the right price

The technology of recycling CFRPs is starting to fall into place, and figures on the energy benefits of recycling carbon fiber seem to add up, according to researchers at Nottingham, who are analyzing the economics of the process (https://www.materialstoday.com/composite-processing/features/new-lease-of-life-for-cfrps/).

"Industry reports claim that the manufacture of rCF achieves about 95% energy reduction compared with vCF, for comparable mechanical performance," points out Meng (www.materialstoday.com).

Overall, rCFRPs can reduce inherent greenhouse gas emissions by about 45% in comparison with vCFRPs, a significant improvement on traditional waste treatment or mechanical recycling.

But no matter how compelling the energy and environmental advantages, the price has to be reasonable—particularly if rCFRPs are going to compete with traditional composites and lightweight materials—for large-scale use across the automotive and other mass-market sectors. The cost of vCF varies depending on the properties, from about $55/kg for standard civil engineering material to $1980/kg for high-end, high-spec fibers for more aerospace applications (Meng et al., 2017; Meng, 2018).

"Our studies, at the University of Nottingham, indicate that the recovery of carbon fibre from CFRP wastes via fluidized bed process can be achieved at under $5/kg with energy savings of over 90%," says Meng. "If large quantities of rCF can be used in future vehicle manufacture, these lightweight structures will achieve benefits of mass reduction, energy saving, GHG emissions reduction, and cost savings."

Regardless of the challenges—in both technological and economic terms—the prospects for recycled CFRPs are appearing attractive. As the first generation of CFRPs come to the end of their useful lives in larger volumes over the next decade, recycling could offer the attractive possibility of an almost closed cycle, whereby recovering carbon fiber from high-performance aerospace composites could drive the adoption of inexpensive rCFRPs in the automotive sector.

"The carbon fibre recycling industry is in its infancy at the moment," says Meng, "but as more recycled carbon fibre becomes available and recycling technology develops, there is the potential to use rCF in higher value products, saving money and being more environmentally sustainable" (www.materialstoday.com).

References

Aircraft Fleet Recycling Association (AFRA), www.afraassociation.org; 2006.
Allen BE. Characterization of reclaimed carbon fibers and their integration into new thermoset polymer matrices via existing composite fabrication techniques. Raleigh, NC: Graduate Faculty of North Carolina State University; 2008 [Thesis for Masters Degree].
Allred RE, Coons AB, Simonson RJ. Proceedings of the 28th SAMPE international technical conference, Seattle, WA, United States, 4–7th November 1996; 1996. p. 11.

Allred RE, Newmeister GC, Doak TJ, Cochran RC, Coons AB. Tertiary recycling of cured composite aircraft parts. Technical Paper EM97-110, Dearborn, MI: Society of Manufacturing Engineers; 1997.

Allred RE, Gosau JM, Shoemaker JM. Proceedings of the 2001 SAMPE symposium & exhibition. Longbeach, CA: SAMPE; 2001.

Anane-Fenin K, Akinlabi E. Recycling of fibre reinforced composites: a review of current technologies. In: Proceedings of the DII—2017 conference on infrastructure development and investment strategies for Africa: infrastructure and sustainable development—Impact of regulatory and institutional framework; 2017.

Anon. Nonwovens Rev 2010;21:3–86.

ATI. Adherent Technologies Inc, www.adherenttech.com; 1994.

Bai Y, Wang Z, Feng L. Chemical recycling of carbon fibers reinforced epoxy resin composites in oxygen in supercritical water. Mater Des 2010;31:999–1002.

Bledzki AK, Kurek K, Barth C. Development of a thermoset partwith SMC reclaim. In: Proceedings of ANTEC 92 50 years: Plastics shaping and the future, technical papers., vol. 50. Detroit, MI: Society of Plastics Engineers; 1992. p. 1558–60.

Butler K. Proceedings of the 46th annual conference, Composites Institute, The Society of Plastics Engineers, Washington, DC, United States, 18–21st February, Session 18B, 1991. 1991. p. 1.

Caliendo H. New bill requests study on carbon fiber recycling. Industry News, Composites World 2015; (22 June 2015), http://www.compositesworld.com/news/new-bill-requests-study-on-carbon-fiber-recycling.

Carberry W. Airplane recycling efforts benefit boeing operators. Boeing AERO Magazine 2008;6 QRT 4.08.

Carberry W. Proceedings of the carbon fibre recycling and reuse 2009 conference. Hamburg, Germany: IntertechPira; 2009.

CFK Valley Stade Recycling GmbH & Co. KG, www.cfk-recycling.com; 2007.

Conroy A, Halliwell S, Reynolds T. Composites recycling in the construction industry. Compos A: Appl Sci Manuf 2006;37(8):1216–22.

Cornacchia G, Galvagno S, Portofino S, Caretto F, Casciaro G, Matera D, Donatelli A, Iovane P, Martino M, Civita R, Coriana S. Carbon fiber recovery from waste composites: an integrated approach for a commercially successful recycling operation. In: Proceedings of the 2009 SAMPE conference. Baltimore, MD: SAMPE; 2009.

Curcuras CN, Flax AM, Graham WD, Hartt GN. Recycling of thermoset automotive components. SAE technical paper series, vol. 910387. 1991. p. 16.

Das S. Life cycle assessment of carbon fiber-reinforced polymer composites. Int J Life Cycle Assess 2011;16:268–82.

Deng J, Xu L, Zhang L, Peng J, Guo S, Liu J, Koppala S. Recycling of carbon fibers from CFRP waste by microwave thermolysis. Processes 2019;7:207.

Eckert CA, Knutson BL, Debenedetti PG. Supercritical fluids as solvents for chemical and materials processing. Nature 1996;383(6598):313–8.

ECRC Services Company, www.ecrcgreenlabel.org; 2003.

EU 2000/53/EC. Directive 2000/53/EC of the European parliament and of the council of 18th September 2000 on end-of-life vehicles. The Council of the European Union, Official Journal of the European Communities; 2000. L 269, GCF, 2010. Green CF.

Feraboli P, Kawakami H, Wade B, Gasco F, DeOto L, Masini A. Recyclability and reutilization of carbon fiber fabric/epoxy composites. J Thermoplast Compos Mater 2012;46:1459–73.

Gardiner G. Recycled carbon fiber update: closing the CFRP lifecycle loop. Composites World 2014; (30 November 2014). http://www.compositesworld.com/articles/recycled-carbon-fiber-update-closing-the-cfrp-lifecycle-loop.

Gosau JM, Wesley TF, Allred RE. Proceedings of the 38th SAMPE technical conference, Dallas, TX, United States, 7–9th November, 2006. 2006.

Goto M. Chemical recycling of plastics using sub and supercritical fluids. J Supercrit Fluids 2009;47:500–7.

HADEG Recycling GmbH, www.hadegrecycling.de; 1995.

Hartt GN, Carey DP. Economics of recycling thermosets. SAE technical paper, vol. 920802. Warrendale, PA: Society of Automobile Engineers; 1992.

Heil JP, Hall MJ, Litzenberger DR, Cleareld R, Cuomo JJ, George PE, Carberry WL. A comparison of chemical, morphological and mechanical properties of various recycled carbon fibers. In: SAMPE'09 conference. Baltimore, MD: SAMPE; 2009.

Hitachi Chemical Co. Ltd, www.hitachichem.co.jp; 2010.

Hyde JR, Lester E, Kingman S, Pickering S, Wong KH. Supercritical propanol, a possible route to composite carbon fibre recovery: a viability study. Compos Part A Appl Sci Manuf 2006;37(11):2171–5.

Janney MA, Newell WL, Geiger E, Baitcher N, Gunder T. Proceedings of the 2009 SAMPE conference. Baltimore, MD: SAMPE; 2009.

Jiang G, Pickering SJ, Lester E, Blood P, Warrior N. Recycling carbon fibre/epoxy resin composites using supercritical propanol. In: Proceedings of the 16th international conference on composite materials, 8–13 July 2007, Kyoto, Japan; 2007.

Jiang G, Pickering SJ, Walker GS, Wong KH, Rudd CD. Surface characterisation of carbon fibre recycled using fluidised bed. Appl Surf Sci 2008;254:2588–93.

Jiang G, Pickering SJ, Lester EH, Turner TA, Wong KH, Warrior NA. Characterisation of carbon fibres recycled from carbon fibre/epoxy resin composites using supercritical n-propanol. Combust Sci Technol 2009;69(2):192–8.

Job S. Recycling composites commercially. Reinf Plast 2014;58:32–8.

Jody BJ, Pomykala JA, Daniels EJ, Greminger JL. A process to recover carbon fibers from polymer-matrix composites in end-of-life vehicles. JOM 2004;56:43–7.

Jutte RB, Graham WD. Proceedings of the 46th annual conference, Composites Institute, The Society of Plastics Engineers, Washington, DC, United States, 18–21st February, Session 18B. 1991. p. 1.

Karborek Spa, www.karborek.it/fibra; 1999.

Khaled DE, Novas N, Gazquez JA, Manzano-Agugliaro F. Microwave dielectric heating: applications on metals processing. Renew Sust Energ Rev 2018;82:2880–92.

Knight CC. Recycling high-performance carbon fiber reinforced polymer composites using sub-critical and supercritical water [PhD thesis]. Florida State University; 2013.

Knight Kouparitsas CE, Kartalis CN, Varelidis PC, Tsenoglou CJ, Papaspyrides CD. Recycling of the fibrous fraction of reinforced thermoset composites. Polym Compos 2002;23(4):682–9.

Knight C, Zeng C, Zhang C and Wang B (2012). Recycling of woven carbon-fibre-reinforced polymer composites using supercritical water Chase Environmental Technology Vol. 33, No. 6, 639–644

Kumar RC, Benal MM, Prasad BD, Krupashankara MS, Kulkarni RS, Siddaligaswamy NH. Microwave assisted extraction of oil from pongamia pinnata seeds. Mater Today Proc 2018;5:2960–4.

La Rosa AD, Banatao DR, Pastine SJ, Latteri A, Cicala G. Recycling treatment of carbon fibre/epoxy composites: materials recovery and characterization and environmental impacts through life cycle assessment. Compos Part B Eng 2016;B104:17–25.

Lester E, Kingman S, Wong KH, Rudd C, Pickering S, Hilal N. Microwave heating as a means for carbon fibre recovery from polymer composites: a technical feasibility study. Mater Res Bull 2004;39(10):1549–56.

Li J, Xu PL, Zhu YK, Ding JP, Xue LX, Wang YZ. A promising strategy for chemical recycling of carbon fiber/thermoset composites: self-accelerating decomposition in a mild oxidative system. Green Chem 2012;14:3260–3.

Liu Y, Liu J, Jiang Z, Tang T. Chemical recycling of carbon fibre reinforced epoxy resin composites in sub-critical water: synergistic effect of phenol and KOH on the decomposition efficiency. Polym Degrad Stab 2012;97:214–20.

Long J, Ulven C, Gutschmidt D, Anderson M, Balo S, Lee M, Vigness J. Recycling carbon fiber composites using microwave irradiation: reinforcement study of the recycled fiber in new composites. J Appl Polym Sci 2015;132(41):42658–65.

Loppinet-Serani A, Aymonier C, Cansell F. Supercritical water for environmental technologies. J Chem Technol Biotechnol 2010;85:583–9.

Markets and markets. Thermoset composites market by manufacturing process (lay-up, filament winding, injection molding, pultrusion), fiber type (glass, carbon), resin type (polyester, epoxy, vinyl ester), end-use industry, and region—Global forecast to 2021, https://www.marketsandmarkets.com/Market-Reports/thermoset-composite-market-140513317.html; 2017.

Marsh G. Reclaiming value from post-use carbon composite. Reinf Plast 2008;52(7):36–9.
Marsh G. Carbon recycling: a soluble problem. Reinf Plast 2009a;53:22–3. 25–27.
Marsh G. Reclaiming value from post-use carbon composite. Reinf Plast 2009b;53(4):22.
Maxime L. Recycling carbon fibre reinforced composites: A market environment assessment, http://digitool.Library.McGill.CA:80/R/-?func=dbin-jump-full&object_id=132167&silo_library=GEN01; 2014.
McConnell VP. Launching the carbon fibre recycling industry. Reinf Plast 2010;54:33–7.
Meng. An assessment of financial viability of recycled carbon fibre in automotive applications. Compos Part A Appl Sci Manuf 2018;109:207–20. https://doi.org/10.1016/j.compositesa.2018.03.011.
Meng F, Mckechnie J, Turner TA. Energy and environmental assessment and reuse of fluidised bed recycled carbon fibres. Compos A: Appl Sci Manuf 2017;100:206–14.
Meyer LO, Schulte K, Grove-Nielsen E. ICCM-16, Japan Society for composite materials, Kyoto, Japan, 2007. 2007.
Morin C, Loppinet-Serani A, Cansell F, Aymonier C. Near- and supercritical solvolysis of carbon fibre reinforced polymers (CFRPs) for recycling carbon fibers as a valuable resource: state of the art. J Supercrit Fluids 2012;66:232–40. https://doi.org/10.1016/j.supflu.2012.02.001.
Nahil MA, Williams PT. Recycling of carbon fibre reinforced polymeric waste. J Anal Appl Pyrolysis 2011;91(1):67–75.
Nakagawa M, Shibata K, Kuriya H. Characterization of CFRP using recovered carbon fibers from waste CFRP. In: Proceedings of the 2nd international symposium on fiber recycling, The fiber recycling 2009 Organizing Committee, Atlanta, GA, United States; 2009.
Oghbaei M, Mirzaee O. Microwave versus conventional sintering: a review of fundamentals, advantages and applications. J Alloys Compd 2010;494:175–89.
Ogi K, Nishikawa T, Okano Y, Taketa I. Mechanical properties of ABS resin reinforced with recycled CFRP. Adv Compos Mater 2007;16(2):181–94.
Okajima I, Sako T. Recycling of carbon fiber-reinforced plastic using supercritical and subcritical fluids. J Mater Cycles Waste Manag 2017;19:15–20.
Okajima I, Watanabe K, Sako T. Chemical recycling of carbon fiber reinforced plastic with supercritical alcohol. J Adv Res Phys 2012;3:1–4.
Oliveux G, Dandy LO, Leeke GA. Current status of recycling of fibre reinforced polymers: review of technologies, reuse and resulting properties. Prog Mater Sci 2015;72:61–99.
Onwudili JA, Yildirir E, Williams PT. Catalytic hydrothermal degradation of carbon reinforced plastic wastes for carbon fibre and chemical feedstock recovery. Waste Biomass Valorization 2013;4:87–93.
Palmer J, Ghita OR, Savage L, Evans KE. Successful closed-loop recycling of thermoset composites. Compos A: Appl Sci Manuf 2009;40:490–8.
PAMELA. PAMELA-life: Main results of the project, www.pamelalife.com/english/results/PAMELA-Life-project_ results-Nov08.pdf; 2008.
Pervaiz M, Panthapulakkal S, Birat KC, Sain M, Tjong J. Emerging trends in automo-tive light- weighting through novel composite materials. Mater Sci Appl 2016;7:26–38.
Pickering SJ. Recycling technologies for thermoset composite materials—current status. Compos A: Appl Sci Manuf 2006;37:1206–15. https://doi.org/10.1016/j.compositesa.2005.05.030.
Pickering SJ. Proceedings of the carbon fibre recycling and reuse 2009 conference. Hamburg, Germany: IntertechPira; 2009.
Pickering SJ, Kelly R, Kennerley JR, Rudd CD, Fenwick NJ. A fluidizedbed process for the recovery of glass fibres from scrap thermoset composites. Combust Sci Technol 2000;60(4):509–23.
Pickering SJ, Liu Z, Turner TA, Wong KH. Applications for carbon fibre recovered from composites. IOP Conf Ser Mater Sci Eng 2016;139:1–18.
Pimenta S, Pinho ST. Recycling carbon fibre reinforced polymers for structural applications: technology review and market outlook. Waste Manag 2011;31(2):378–92. https://doi.org/10.1016/j.wasman.2010.09.019.
Pimenta S, Pinho ST. Recycling of Carbon Fibers. In: Handbook of recycling; 2014.
Piñero-Hernanz R, García Serna J, Dodds C, Hyde J, Poliakoff M, Cocero MJ, Kingman S, Pickering S, Lester E. Chemical recycling of carbon fibre reinforced composites in near critical and supercritical water. Compos A: Appl Sci Manuf 2008;39(3):454–61.

Princaud M, Aymonier C, Loppinet-Serani A, Perry N, Sonnemann G. Environmental feasibility of the recycling of carbon fibers from CFRPs by solvolysis using supercritical water. ACS Sustain Chem Eng 2014;2:1498–502.

Roberts T. The carbon fibre industry: Global strategic market evaluation 2006–2010. 10. Watford, Hertfordshire: Materials Technology Publications; 2006.

Roberts T. Rapid growth forecast for carbon fibre market. Reinf Plast 2007;51:10–3.

Rybakov KI, Buyanova MN. Microwave resonant sintering of powder metals. Scr Mater 2018;149:108–11.

Scheirs J. Polymer recycling, science, technology and applications. Chichester, UK: John Wiley & Sons; 1998.

Sims B, Booth CA. Inventors; Phoenix Fibreglass, Inc., assignee; International Patent WO 93/05883;1993.

Sloan J. Proceedings of the high performance composites carbon fiber 2007 conference coverage, Composites World, Yarmouth, ME, United States, 5–7th December, 2007; 2007.

Soh SK, Lee DK, Cho Q, Rag Q. Low temperature pyrolysis of SMC scrap. In: Proceedings of 10th annual ASM/ESD advanced composites conference, Dearborn, Michigan, United States, 7–10 November 1994; 1994. p. 47–52.

Spooner M (2012). Private communication.

Sugeta T, Nagaoka S, Otake K, Sako T. Decomposition of fiber reinforced plastics using fluid at high temperature and pressure. Kobunshi Ronbunshu 2001;58:557–63 [in Japanese].

Sun H, Guo G, Memon SA, Xu W, Zhang Q, Zhu JH, Xing F. Recycling of carbon fibers from carbon fiber reinforced polymer using electrochemical method. Compos Part A Appl Sci Manuf 2015;78:10–7.

Takahashi J, Matsutsuka N, Okazumi T, Uzawa K, Ohsawa I, Yamaguchi K. Mechanical properties of recycled CFRP by injection molding method. In: 16TH International conference on composite materials, Japan Society for Composite Materials, Kyoto, Japan; 2007.

Ushikoshi K, Komatsu N, Sugino M. Recycling of CFRP by pyrolysis method. J Soc Mater Sci 1995; 44(499):428–31.

Wang Y, Cui X, Ge H, Yang Y, Wang Y, Zhang C, Hou X, Li J, Deng T, Qin Z. Chemical recycling of carbon fiberreinforced epoxy resin composites via selective cleavage of the carbon − nitrogen bond. ACS Sustain Chem Eng 2015;3(12):3332–7.

Warrior NA, Turner TA, Pickering SJ. AFRECAR and HIRECAR project results. In: Carbon fibre recycling and reuse 2009 conference. Hamburg, Germany: IntertechPira; 2009.

Witik RA, Teuscher R, Michaud V, Ludwig C, Manson JAE. Carbon fibre reinforced composite waste: an environmental assessment of recycling, energy recovery and landfilling. Compos A: Appl Sci Manuf 2013;49:89–99. https://doi.org/10.1016/j.compositesa.2013.02.009.

Wong KH, Pickering SJ, Turner TA, Warrior NA. Compression moulding of a recycled carbon fibre reinforced epoxy composite. In: SAMPE'09 conference. Baltimore, MD: SAMPE; 2009.

Wong K, Rudd C, Pickering S, Liu X. Composites recycling solutions for the aviation industry citation. Sci China Technol Sci 2017;60:1291. https://doi.org/10.1007/s11431-016-9028-7.

Xu P, Li J, Ding J. Chemical recycling of carbon fibre/epoxy composites in a mixed solution of peroxide hydrogen and N,N-dimethylformamide. Combust Sci Technol 2013;82:54–9. https://doi.org/10.1016/j.compscitech.2013.04.002.

Yadoji P, Peelamedu R, Agrawal D, Roy R. Microwave sintering of ni–zn ferrites: comparison with conventional sintering. Mater Sci Eng B 2003;98:269–78.

Yan H, Lu C, Jing D, Chang C, Liu N, Hou X. Recycling of carbon fibers in epoxy resin composites using supercritical 1-propanol. New Carbon Mater 2016;31(1):46–54.

Yip HLH, Pickering SJ, Rudd CD. Characterisation of carbon fibres recycled from scrap composites using fluidised bed process. Plast Rubber Compos 2002;31(6):278–82.

Yuyan L. Recycling of carbon/epoxy composites. J Appl Polym Sci 2004;94(5):1912.

Yuyan L, Guohua S, Linghui M. Recycling of carbon fibre reinforced composites using water in subcritical conditions. Mater Sci Eng A 2009;520:179–83.

Zhu JH, Pi-yu C, Su M, Peia C, Xing F. Recycling of carbon fibre reinforced plastics by electrically driven heterogeneous catalytic degradation of epoxy resin. Green Chem 2019;21:1635–47.

Relevant websites

https://www.compositesworld.com/articles/carbon-fiber-life-beyond-the-landfill.
https://www.materialstoday.com/composite-processing/features/new-lease-of-life-for-cfrps/.
https://www.press.bmwgroup.com/global/startpage.html, https://www.sglgroup.com/cms/international/products/product-groups/cf/recafil/index.html?__locale=en.
pubs.rsc.org.
pure.ltu.se.
www.cfk-recycling.com.
www.compositesworld.com.
www.hadegrecycling.de.
www.karborek.it/fibra;.
www.materialstoday.com.
www.reinforcedplastics.com.

Further reading

BMW Press Club Global, https://www.press.bmwgroup.com/global/startpage.html; 2015.
ERCOM n.d. Composite recycling—Raw material for the future, ERCOM Composite Recycling GmbH, Germany.
Gosau JM (2012). Private communication.
Recycomp. Recycling technology for composites based on molten salts, Paris, France: Omnexus; 2006. http://www.omnexus.com/resources/innovation/news.
SGL. RECAFIL® recycled carbon fibers, https://www.sglgroup.com/cms/international/products/product-groups/cf/recafil/index.html?__locale=en.

CHAPTER 7

Manufacturing of composites from recycled carbon fiber

Based on single-filament tensile data, recycled carbon fiber (RCF) can supplement or replace glass and virgin carbon fiber (VCF) under appropriate conditions. RCF that retains 80%–90% of virgin fiber strength at gauge lengths of ≥ 25 cm is an attractive replacement for glass fiber. RCF offers comparable mechanical properties in a lighter material for a nominal price increase compared with E-glass. Work by Heil et al. (2009) showed that recycled intermediate modulus carbon fiber is stronger than virgin-standard modulus carbon fiber. RCF shows reduction of a few percent or less in stiffness compared with VCF; another possible use of RCF is in stiffness-driven applications in which a slight reduction in strength is acceptable. However, two challenges prevent RCF from being readily adopted as a structural material (Pimenta and Pinho, 2011).

The first challenge is integrating RCF into existing composite manufacturing methods. RCF is discontinuous, short, and often tangled. Hence, feeding RCF into a molding process in a reliable and controlled manner must be realized.

The second challenge is that data from single-filament testing have limited usefulness in traditional computer-aided design/finite element method (FEM) product design. Researchers from Imperial College London (London, United Kingdom) have shown that strain energy release rate fracture toughness can be modeled in a RCF laminate based on single-filament mechanical data and the architecture of the reinforcement fabric.

Pimenta and Pinho (2011) reviewed the manufacture of composites from RCF. "The reclaimed fibres are steeped with a new matrix in the second stage of the Carbon Fiber reclamation processes. The RCF is generally fragmented into shorter lengths as a result of the size reduction of carbon fibre-reinforced polmer (CFRP) waste before reclamation, breakage of fibre during reclamation, and chopping of fibres after reclamation. Furthermore, all fibre reclamation processes remove the sizing from the fibres, therefore the recyclate is in a filamentised, random, low-density packing form. So, the manufacturing processes developed for virgin materials (generally available as sized tows) should be adapted to the unique recycled-fibre form" (Pimenta and Pinho, 2011).

Several methods are used to produce composites using recovered carbon fiber (Pimenta and Pinho, 2011):
- direct molding,
- bulk molding compound compression,
- compression molding of nonwoven products,
- fiber alignment,
- woven recycled carbon fiber-reinforced polymer (RCFRP).

There are two direct methods of remolding recovered carbon fiber into recycled composite materials (Astrom, 1997):
- injection molding
- press molding of bulk molding compounds (BMC)

In the injection molding the material is introduced into a heated barrel, mixed, and forced into a mold cavity where it gets cooled and hardened to the configuration of the mold cavity. BMC combines short fibers and resin in the form of a bulk prepreg. BMC is well suited for compression or injection molding (pure.ltu.se).

Wong et al. (2007) used injection molding for manufacturing a polypropylene composite reinforced with RCF. With this process, fibers aligned in one direction, which resulted in high stiffness of the composite material in that direction. With the addition of Maleic anhydride-modified polypropylene (MAPP), there was improvement in the interfacial adhesion as fewer fibers were pulled out, and lesser debonding was observed. This produced composites with higher tensile and flexural strengths. The maximum strength was obtained from MAPP with the highest molecular weight. With certain grades of MAPP, increased modulus was obtained. The composite impact strength improved substantially by MAPP, because of a higher compatibility between the fiber and matrix, which reduced crack initiation and propagation.

A molding compound process was developed by Turner et al. (2009). This process involved sheet molding compounds (SMC) and BMC. The formulation of the BMC was tuned for overcoming the poor flow properties of the resin and the entangled form of the fibers.

Connor (2008) compared the performance of two injected CFRP: one with virgin carbon fiber and other with RCF (from Recycled Carbon Fibre Ltd). The recyclate was 25% less stiff in comparison with the virgin control. The strength reduction was less only 12%, which could be because of improved fiber-matrix adhesion in the recyclate. With fibers from ATI (Adherent Technologies Inc, 1994), the same process was not successful due to their more dispersed structure and poor adhesion of fiber-matrix. BMC are intermediate products produced by mixing RCF, resin, and fillers and curing agents into bulky charges. This mixture is then compression molded under 3:5 MPa to 35 MPa into a component (Astrom, 1997; Allen, 2008). Fibers from adherent and milled carbon have been compounded into injection molding and BMCs for assessment. Performance characteristics were better as compared with those of glass-reinforced materials and, in the case of injection molding, were competitive with off-the-shelf virgin carbon fiber-filled compounds (Aircraft and Composite Recycling, 2007). RCF F18 has also been incorporated directly into fiber preforms for a compression-molding demonstration. Materials Innovation Technologies (Fletcher, North Carolina, United States) successfully manufactured preforms directly from as provided fiber after chopping and had them molded into a production configuration-automotive component. This Corvette C6 fender-wall component manufactured from RCF is lighter by about 20% in comparison with the production fiberglass component even without engineering for improved stiffness (Aircraft and Composite Recycling, 2007).

Carbon/epoxy composites have been manufactured using high-quality CF recovered using fluidized bed and thermo/chemical processes (Pickering, 2006; Turner et al., 2009). BMC have been manufactured by the inclusion of short (6–12 mm) carbon filaments directly into filled epoxy matrices at fiber volume fractions of 10%–15%. CF SMC and prepregs have been manufactured using a partially aligned intermediate mat material. Mats have been manufactured using a papermaking process and impregnated with filled and unfilled epoxy resin films. In the direction of alignment, Young's modulus of $\leq$80 GPa and tensile strength of 425 MPa have been observed for high volume fraction (45%) press-molded laminates.

Materials have also been developed for molding under vacuum only and autoclave conditions and for applications requiring flow during molding. Materials have been shown to outperform standard glass polyester-molding compounds at a fraction of the cost of emerging virgin CF-based epoxy compounds.

The major factors that affect the mechanical performance of the RCFRP (particularly strength) were the fractions of fillers and of RCF. The mechanical performance of the RCFRP was found to be better in comparison with that of commercial glass BMC (Warrior et al., 2009). It is not clear whether these RCFRP can compete in price (Pickering et al., 2010).

Another method for manufacturing composite materials with longer fibers using recovered carbon fiber is to produce intermediate dry two-dimensional (2-D) or three-dimensional (3-D) nonwoven performs (Pimenta and Pinho, 2011; Wong et al., 2007; Pickering, 2006; Janney et al., 2009). "The 2D or 3D non-woven dry products are then subjected to compression moulding with resin layers or re-impregnated through a liquid resin process. In a RCF, a SMC was produced using an intermediate mat material made of Carbon Fiber recovered using a fluidised bed process. These composites were produced using a papermaking process and interleaved with epoxy resin film. Fibres recovered by using pyrolysis and converted into a randomly oriented mat were used for manufacturing a composite material using a resin film infusion process in a matched die tool. The characterisation of this composite was done in terms of its microstructure and mechanical performance" (pure.ltu.se).

Janney and Baitche (2007) introduced the 3-D engineered preform (3-DEP) process for producing a highly uniform, complex geometry, net shape-chopped fiber preforms for polymer matrix composites. This process was developed using virgin fibers by Materials Innovation Technologies (www.nemergingmit.com) and used by Janney and Baitche (2007) for producing preforms for high-quality composite parts using RCF. Reuse of recovered carbon fiber has been limited to mainly new thermoset composite materials using compression molding or resin film infusion (Jiang et al., 2006a,b; Pimenta et al., 2009, 2010; Thompson et al., 1998). Research was conducted by Szpieg et al. (2012) in which the recovered carbon fiber was introduced in a recycled thermoplastic matrix by press forming.

Recovered carbon fibers are usually highly entangled, so many methods have been used for achieving a desired uniformity of the fiber distribution before reprocessing them into new composites (Yip et al., 2002; Meyer et al., 2009).

A method for dispersing recovered carbon fiber has been developed by Szpieg et al. (2012). The used fibers were recovered using a pyrolysis process. These fibers were delivered in a very fluffy structure not appropriate for use in new composites. Thanks to this method, uniform fiber distribution was developed. Here the principles of papermaking were used and found to be appropriate for fiber dispersion. In this process, fibers were dispersed in distilled water using a mixer. After dispersion a filter was used for draining off the water and protecting the fiber preform from damage. The carbon fiber preform was dried before further processing of composite. The resulting carbon fiber preform was used for producing thermoplastic composite material by press forming.

A further method for dispersing recovered carbon fiber is to process their fluffy form and align them into a high-performance mat (Jiang et al., 2006a,b). Fibers were recovered using a fluidized bed process. "For achieving alignment, fibre suspension in a cellulose solution was subjected to a velocity gradient by using an alignment box. The fibres in the slurry film were aligned in the direction of flow. The mat was formed by feeding the slurry film onto the inside of a rotating permeable cylindrical drum moving with an angular velocity enough to pass the carrier medium swiftly through the surface by centrifugal force" (Jiang et al., 2006a,b).

Fiber alignment is a critical factor for improving the properties from recovered fibers. It improves the mechanical performance of composites produced with discontinuous RCF (Pickering, 2009). Also the mechanical properties of the composite improve along the preferential fiber direction as manufacturing needs lower molding pressures and smoother fiber-to-fiber interactions (Wong et al., 2007; Turner et al., 2009). The modified papermaking method developed by Pickering (2009), Turner et al. (2009), and Warrior et al. (2009), in collaboration with Howarth and Jeschke (2009) and Technical Fibre Products (www.techfibres.com), is a proprietary process for random nonwoven 2-D mats. Presently, up to 80% of the theoretical upper directional alignment is reached, using shorter RCF and thin mats (down to 10 gsm). This method produced RCFRP with the highest mechanical properties ever. But the filamentized RCF form reduced the impact energy to half of that usually measured for glass fiber-reinforced polymer (GFRP) SMC (Turner et al., 2009). Presently, research is focusing to improve the packability of mats and through the thickness uniformity of alignment (Wong et al., 2007; Turner et al., 2009).

The 3-DEP process developed by Janney and Baitche (2007) was used for producing a virgin carbon fiber-reinforced polymer (VCFRP) cone with fibers preferentially aligned circumferentially; this was obtained by adjusting the position and motion of the deposition tool. A centrifugal alignment rig was presented by Wong et al. (2007). It uses a rotating drum equipped with a convergent nozzle, aligning a suspension of RCF that is highly dispersed. The use of shorter fibers improved the RCFRP alignment obtained ($\leq$90%). A yarn-spinning method has been developed by Wong et al. (2007), within the FibreCycle project. Wet dispersions of RCF are passed using a pipe with an induced vortex. Spun yarns with 50 filaments and 60 mm long are produced under the optimized conditions.

Liu et al. (2012) studied the combination of phenol and potassium hydroxide for chemically recycling carbon fiber-reinforced epoxy resin cured with 4,4′-diaminodiphenylmethane in subcritical water. This combination showed a synergistic effect on decomposing this type of epoxy resin. Compared with virgin carbon fiber after sizing removal, the surface compositions of the recovered carbon fiber showed little change, and the tensile strength of the recovered carbon fiber was found to retain very well.

Heil et al. (2011) used a wet-lay process to make lightweight, randomly oriented fiber mat preforms. Fiber preforms were made by blending standard modulus and intermediate modulus CF of virgin and recycled quality. Analyses of RCF showed 80%–90% strength retention at 13-mm gauge length compared with virgin carbon fiber. Composite panels molded from RCF offer a cost-competitive alternative to SMC made with glass fiber while maintaining comparable mechanical properties.

Carbon fibers recovered by thermochemical methods are being considered for reusing in new composite materials (Table 7.1). Random and aligned discontinuous fibers have been studied (Elghazzaoui, 2012; Feraboli et al., 2007; Boeing, 2007; Connor, 2008; Wong et al., 2010, 2012; Heil and Cuomo, 2011; Szpieg, 2011; Akonda et al., 2012; Meredith et al., 2012; Pimenta, 2013; Stoeffler et al., 2013; Illing-Günther et al., 2013), but very few real prototypes have been produced (Table 7.2).

Furthermore, two types of rCF can be identified: random short fibres from either woven or nonwoven fibre composites and pieces of woven fabrics. The size of recovered pieces of woven fabrics depends on the shredding pre-treatment; however the fabrics can actually retain their woven shape after the recycling treatment. This can be very interesting in terms of fibre alignment

Table 7.1 Composite materials manufactured with rCF for mechanical testing.

North Carolina State University (Connor, 2008; Heil and Cuomo, 2011)	Injection molded compounds with rCF from ELGCF and ATI and polycarbonate, rCF-reinforced plastics by RTM; fiber mats then a preform. The wet-lay process, commonly used for making paper
Université du Quebec Montreal, Bell Helicopter Textron Canada and NRCC, Canada (Stoeffler et al., 2013)	rCF mats, then used to manufacture infusion/compression molded composites Pellets of PPS reinforced with rCF molded by injection
FibreCycle project, United Kingdom (Akonda et al., 2012)	rCFs (length > 500 mm) blended with thermoplastic staple fibers into yarns
University of Nottingham (Wong et al., 2012)	CF veils to manufacture electromagnetic interference shielding by RTM PP/rCF by compounding and injection molding, with and without coupling agents
National Technical University of Athens, Greece (Kouparitsas et al., 2002)	Ground CFR epoxy in ionomer Ground GFR polyester in polypropylene Ground aramid fiber-rereinforced epoxy in ionomer
University of Warwick, Umeco Composites Structural Materials, Lola group and ELG Carbon Fiber (Pimenta, 2013)	UK Woven CFs reused in high-performance energy absorption structures University of Warwick: reuse of woven rCF to manufacture the rear of WorldFirst F3 car
Ehime University and Toray Industries Inc., Japan (Ogi et al., 2007) University of Tokyo and Toray Industries Inc., Japan (Takahashi et al., 2007)	ABS reinforced with crushed CFRC by injection molding (fiber length about 200 lm) Japan ABS and PP reinforced with crushed CFRC by injection (CFRC square flakes of about 1 cm)
Imperial College of London, United Kingdom (Pimenta, 2013)	rCF recovered from uncured CF/epoxy fabrics by ELGCF pyrolysis and reused in a laminate made by resin film infusion (RFI)
Luleå University of Technology and Swerea SICOMP (Szpieg, 2011)	Sweden rCF reused in preforms (mat, by papermaking method) then laid up with films of PP scrap and hot press formed
Saxon Textile Research Institute, Germany (Illing-Günther et al., 2013)	Nonwoven rCF fabrics (stich bonded or needle punched) suitable for medium-strength requiring applications
University of Seattle, Automobili Lamborghini and Japan Ministry of Defense, United States, Italy and Japan (Feraboli et al., 2012)	Entangled rCF randomly distributed recovered from woven fabrics, however, with a majority of fibers still in tow shape. The initial length of the fibers is retained. Remanufacturing by direct incorporation and VaRTM
University of Nantes, ICAM and Compositec, France (Elghazzaoui, 2012)	Realigned rCF to manufacture UD coupons BMC and SMC
University of Ulster, United Kingdom (Archer et al., 2009)	Polymer composite materials made up of waste CFs (collected from a 3-D weaving process) and manufactured by injection and compression molding

(Based on Oliveux G, Dandy LO, Leeke GA, Current status of recycling of fibre reinforced polymers: review of technologies, reuse and resulting properties. Prog Mater Sci 2015;72:61–99.)

Table 7.2 Real part demonstrator and commercially available semiproducts made of rCF.

Company/organization	Products	
Materials Innovation Technologies RCF (MIT-RCF), United States http://www.boeingsuppliers.com/environmental/TechNotes/TNdec07.pdf	Preforms and finished parts by the 3-DEP process (e.g., front lower wheelhouse support for the Corvette)	Real part demonstrator
Janicki Industries and Boeing, United States Pimenta, 2013	A tool for composite layup	
Technical Fibre Products (TFP), United Kingdom http://www.tfpglobal.com/materials/carbon	Carbon veils (Optiveil eco) and mats (Optimat eco)	Commercially available
Sigmatex Ltd., United Kingdom http://www.sigmatex.com/Solution/Recycled_Fabrics	Commingled thermoplastic fabric based on rCF and available PET matrix	

(Based on Oliveux G, Dandy LO, Leeke GA, Current status of recycling of fibre reinforced polymers: review of technologies, reuse and resulting properties. Prog Mater Sci 2015;72:61–99.)

and so in terms of reinforcement retention. This type of reinforcement has received little attention whereas it represents more than 60% of Carbon Fiber waste just in Europe.

(Feraboli et al., 2007; Connor, 2008; Meredith et al., 2012)

Most often rCFs are not woven; in a wide range of lengths the fibres are intermingled like a tuft of hair. Different trials have been performed using rCF either as is or reshaped by different techniques (wet papermaking method, dispersion in liquid medium or by compressed air). The other important issue is the absence of sizing on the recycled fibre surface and the presence of residues that can coat the fibres. Due to their short length, rCFs were first incorporated into random discontinuous fibre materials like SMCs and BMCs. They generally need to be chopped again after reclamation in order to conform to virgin fibre length commonly used for this type of material.

(Connor, 2008; Pimenta, 2013)

Some studies used rCFs as they were but the resulting materials were not homogenous in fibre distribution, thus in thickness, resulting in a quite low fibre content and high void content. Inspite of the fibre content lower in comparison to that of a BMC with vCF, the flexural properties were comparable to the virgin BMC. Ultimate tensile strength reduced by 5% and flexural modulus reduced by 15%. The flexural properties obtained for an SMC were between 87% and 98% of those obtained for the SMC with vCFs, fibre contents were comparable.

(Iwaya et al., 2008; Yuyan et al., 2009; Elghazzaoui, 2012; pure-oai.bham.ac.uk)

Other methods depend on the material to recycle and especially on the reuse of the fibers. The recovered products, mostly fibers and products from resin decomposition, are mostly systematically characterized and show that they can be reused. Particularly, rCFs (often discontinuous as the materials need to be cut in smaller pieces before treatment) have been incorporated in a few studies successfully (Molnar, 1995; Palmer et al., 2010; Elghazzaoui, 2012; Boeing, 2007; Janney and Baitche, 2007; Connor, 2008.

Wong et al., 2010; Heil and Cuomo, 2011; Szpieg, 2011; Wong et al., 2012; Akonda et al., 2012; Meredith et al., 2012; Pimenta, 2013; Stoeffler et al., 2013; Illing-Günther et al., 2013; Archer et al., 2009).

> *On the other hand, fractions containing products from resin degradation by solvolysis have received very little consideration. Pyrolysis products from the resin have been mainly considered as a source of energy to feed back into the process. Currently solutions do exist to recycle composite materials. It can be seen in the literature that many different processes and methods have been applied and have shown the feasibility of recycling such materials, some of them being more commercially mature than others. However, industrial applications using recycled fibres or resins are still rare, partly because of a lack of confidence in performance of rCFs, which are considered as of lower quality than virgin carbon fibres (vCFsbut also because rCFs are not completely controlled in terms of length, length distribution, surface quality (adhesion to a new matrix) or origin (often different grades of fibres are found in a batch of recycled composites coming from different manufacturers). Furthermore, recycling of composites is globally not closed-loop in terms of resource efficiency as recycled fibres cannot be reused in the same applications as their origin. For example if they come from an aircraft structural part, they cannot be reused in a similar part. In light of this, relevant applications have to be developed and specific standards are required in order to manage those reclaimed fibres. Demonstrators have been manufactured with rCFs, showing potential new applications.*

(Pimenta and Pinho, 2011; McConnell, 2010)

The feasibility of using the high-performance discontinuous fiber (HiPerDiF) method for producing highly aligned discontinuous fibers intermingled hybrid composites with flax and rCF, and the potential benefits of doing so was studied by Longana et al. (2018). "HiPerDiF technology was used to manufacture ADFRC with different ratios of rCF and flax fibres and their mechanical, i.e., stiffness and strength, and the functional properties, i.e., vibration damping, were evaluated through tensile tests and the half-power bandwidth method, 100% rCF ADFRC manufactured with the HiPerDiF technology display high primary mechanical properties, i.e., stiffness and strength. However, it is interesting to notice the effect of flax fibres addition. Flax fibre price is 10% of the rCF one, therefore, even a small addition allows to sensibly reduce the cost of the finite composite as well as its weight. This happens at the expense of the primary mechanical properties, i.e., stiffness and strength. However, the functional benefits of intermingled hybrids are evident: an increase of 2.5 times in the damping properties can be achieved if 75% of the rCF are substituted with flax fibres. Flax/rCF intermingled hybrid materials manufactured with the HiPerDiF method can be a viable solution in applications where a reduction of primary mechanical properties is an acceptable trade-off for the increase of functional properties and cost reduction" (www.mdpi.com).

Some recycling processes are able to preserve the reinforcement architecture of the waste, so it is possible to recover the structured weave from large woven items (example out-of-date prepreg rolls, end-of-life aircraft fuselage, or prepreg trimmings from large components). Reimpregnation of the recycled weave fabrics then produces woven RCFRP through, for instance, resin transfer molding or resin infusion. With presently available recycling processes, stiffness and strength can reach more than 70 GPa and 700 MPa, respectively. Furthermore, fabrics recovered from prepreg rolls would be fully traceable.

Allen (2008) used woven fabrics from undisclosed recyclers; the mechanical properties of the RCFRP were poor when compared with similar VCFRP (particularly tensile strength), because of degradation of fiber during recycling.

Meredith (2009) applied woven RCFRP to noncritical parts of an environmentally sustainable Formula-3 car. The car also used other recycled and natural materials and biofuels.

George (2009) fabricated a RCFRP tool (https://www.janicki.com) for composite layup in aircraft manufacturing.

Jiang et al. (2006a,b) used a hydrodynamic alignment method to process the fluffy form of the RCF into a high-performance aligned mat. The alignment of the fiber was characterized by producing the aligned mat into epoxy resin composite. Optical microscopy of the cross section of the composite and the ratio of the mechanical properties along and transverse the fiber direction indicated a good degree of fiber alignment. It was observed that shorter fibers, 5 mm in length, can be aligned more easily than longer fibers of 20 mm. The research revealed that carbon fiber, produced as fluffy, short filaments in their recycled form, could be upgraded for producing an aligned product suitable for higher-grade applications.

Meredith et al. (2012) made a comparison of the mechanical properties of virgin and recycled woven carbon fiber prepregs and assessed the potential for RCFRP for using in high-performance energy-absorption structures. "Three sets of material were examined: 1) fresh containing virgin fibres; 2) resin, aged which was an out-of-life but otherwise identical roll; and 3) recycled, which contained recycled fibre and new resin. The compressive strength and modulus of RCFRP were ≈94% of the values for fresh material. This correlated directly with the results from impact testing where RCFRP conical impact structures were found to have specific energy absorption of 32.7 kJ/kg compared with 34.8 kJ/kg for fresh material. The tensile and flexural strength of RCFRP were 65% of the value for fresh material. Tensile and flexural moduli of RCFRP were within 90% of that for fresh material and the interlaminar shear strength of RCFRP was 75% that of fresh material. Overall, RCFRP has been shown to remain a highly satisfactory engineering material. This is an important finding because it proves that Carbon Fiber can be recycled and reused in high-performance applications. This has significant implications for the use of Carbon Fiber in the automotive industry where End-of-life Vehicle legislation requires 85% of materials to be recyclable. This work opens the door for the significant expansion of use of Carbon Fiber in the automotive industry. Of equal important is the finding that the mechanical properties of fresh and aged Carbon Fiber were all within 6% of one another. This highlights an opportunity to understand resin systems in more detail to prevent the creation of composite waste at the source."

Akonda et al. (2012) produced comingled composite fiber/polypropylene (CF/PP) yarns from chopped RCF (length, 20 mm; diameter, 7–8 μm) blended with matrix PP staple fibers (length, 60 mm; diameter, 28 μm) using a modified carding and wrap-spinning process. "Microscopic analyses showed that more than 90% of the chopped RCF was aligned along the yarn axis. Thermoplastic composite test specimens fabricated from the wrap-spun yarns had 15–27.7% chopped RCF volume content. Similar to the yarn, more than 90% of the chopped RCF comprising each composite sample made showed parallel alignment with the axis of the test specimens. The average values obtained for tensile, and flexural strengths were 160 MPa and 154 MPa, respectively, for composite specimens containing 27.7% chopped RCF by volume. It was concluded that, with such mechanical properties, thermoplastic composites made from RCF could be used as low-cost materials for many non-structural applications."

Members of a UK consortium—Advanced Composites Group (part of Umeco Composites Structural Materials), Exel Composites (United Kingdom), Net Composites, Sigmatex, Tilsatec, and the University of Leeds—have launched a new family of yarns and fabrics produced from RCF that offers properties rivaling those of virgin carbon fiber but costing less (Thryft, 2010).

> The materials are the result of the consortium's recently concluded FibreCycle collaborative project, which developed technology for recovering waste Carbon Fiber, blending it with resins such as polyethylene terephthalate (PET), and converting it into different types of materials. It recovered Carbon Fiber from all stages of manufacturing, and developed reprocessing methods within a representative supply chain. The new Carbon Fiber materials are continuous, highly aligned reinforcements (including yarns, woven textiles, non-crimp fabrics, prepreg tapes, and preconsolidated sheets). Like other commingled and blended materials, the fabrics are moulded under pressure and pass through a heating and cooling cycle. Several demonstrator products (including press-molded automotive parts) have been produced. The commercial members of the consortium serve multiple industries (including automotive, aerospace, and industrial). Composite laminates made from 50% recycled carbon and 50% recycled PET, measured by weight, can deliver ≥90% of the tensile modulus, or stiffness, and 50% of the tensile strength of an equivalent composite made from virgin fibres.
>
> **(https://www.arabplasticsnews.com)**

The properties obtained show that the materials are suitable for several applications, particularly in the automotive, aerospace, sports and leisure, medical, and energy sectors. Though these new fibers and fabrics have almost similar stiffness as virgin materials, the carbon fiber feedstock on which they are based costs less than virgin fibers and fabrics would (https://myplasticsblog.wordpress.com).

Takayasu KK have developed Takayasu Recycled Carbon Fiber (TRCF) made from unused CFRP and prepreg (Anon, 2010). Optimization of pyrolysis conditions that decompose plastics (but not CF) is the key to this technology. This enables the recovery of carbon fiber in staple form, which is a clear difference from those in powder form that are recovered from CFRP and prepreg by conventional methods. Also, TRCF retains 90% of the mechanical strength of the original carbon fiber, and TRCF-nonwoven sheets show excellent conductivity and shielding performance. Takayasu KK have also focused on highly functional polyester staples suitable for nonwoven production. The antibacterial type uses an organic antibacterial agent that is resistant to high temperatures and is particularly effective in eliminating various types of fungi. The super water-repellent type can be blended into other types of fibers to make water-repellent nonwovens. The flame-retardant type shows good color fastness to light and is ideal as interior material for cars. The potential for processing recovered carbon fiber into nonwovens was investigated by Hofmann and Schreiber (2012). Carbon fiber-reinforced plastics are strong, lightweight materials that are used mainly in aviation, in the generation of wind energy and, increasingly, in the automotive industry. There is little in the way of environmentally efficient recycling of carbon fiber; most carbon fiber waste is used for energy-recovery purposes. Processing recovered carbon fiber into nonwovens represents a form of material recycling that can make the best use of the inherent properties of carbon fiber. Investigations revealed that it is possible to use mechanical carding to form webs made of 100% primary carbon fiber and RCF. This technology offers new potential for the industry to process carbon fiber waste, for the first time, in a large-scale and economical fashion without compromising the high quality of the carbon fiber. Research into using the same technology to process RCF from the pyrolysis of disused carbon fiber components is in progress.

Research on RCFRP manufacturing is in progress (Pimenta and Pinho, 2011). Reimpregnating nonwoven mats is one of the most efficient methods in terms of the mechanical performance of the composites (Warrior et al., 2009; Janney et al., 2009; Wong et al., 2009). Properties are at the level of structural virgin materials (e.g., GFRP, short-fiber CFRP, and aluminum). More research is required in remanufacturing technologies, particularly concerning inducing fiber alignment, increasing fiber content, and reducing fiber damage during processing (Pimenta et al., 2010; Pickering, 2009; Pickering et al., 2010). The performance of most RCF is similar to that of virgin fibers, so recycled composites could (in principle) reach the properties of VCFRP with comparable architectures if appropriate remanufacturing processes were developed. A few structural components have been produced with RCFRP: secondary components for the automotive industry, components for aircraft interiors, and tool. Pimenta and Pinho (2011) have summarized the benefits and problems with different processes used for remanufacturing of composites from RCF (Table 7.3).

Table 7.3 Benefits and problems with different processes used for remanufacturing of composites from recycled carbon fibers.

Direct molding
Benefits: processes already established, mechanical performance compatible with low- or medium-end structural applications.
Problems: very low fiber contents, reduced fiber length, difficult processing due to rCFs' filamentized form.
Compression molding of nonwoven products
Benefits: processes requiring minor adaptations only, processes widely used for rCFRP and well documented, mechanical properties comparable with virgin structural materials, potential application in automotive and aircraft industries.
Problems: common fiber damage during compression molding, competing market dominated by relatively cheap materials.
Compression molding of aligned mats
Benefits: improved uniaxial mechanical properties, possibility of tailoring the layup of rCFRP laminates, potential for preserving fiber length and achieving higher reinforcement fractions.
Problems: need for nearly perfect alignment to significantly improve packability, need for substantial development of processes.
Impregnation of woven drawback
Benefits: structured architecture with continuous fibers and high reinforcement content, simplicity of manufacturing processes, applicable demonstrators already manufactured.
Problems: applicability currently reduced to prepreg EoL rolls, experimental realization of theoretical mechanical properties still to be measured.

(Based on Pimenta S, Pinho ST, Recycling carbon fibre reinforced polymers for structural applications: technology review and market outlook. Waste Manag 2011;31:378–92.)

References

Adherent Technologies Inc. www.adherenttech.com; 1994.
Aircraft and Composite Recycling. http://www.boeingsuppliers.com/environmental/TechNotes/TNdec07.pdf; 2007.
Akonda MH, Lawrence CA, Weager BM. Recycled carbon fibre-reinforced polypropylene thermoplastic composites. Compos A: Appl Sci Manuf 2012;43:79–86.
Allen BE. Characterization of reclaimed carbon fibre and their integration into new thermoset polymer matrices via existing composite fabrication techniques. Raleigh, NC: North Carolina State University; 2008 [thesis for masters degree].
Anon. Nonwovens Rev 2010;21(3):86.
Archer E, McIlhagger AT, Buchanan S, Dixon D. Reuse of waste carbon fibre by compounding with thermoplastic polymers. In: Proceedings of the 17th international conference on composite materials (ICCM-17), 27–31 July 2009, Edinburgh, UK. 2009.
Astrom BT. In manufacturing of polymer composites. London: Chapman and Hall; 1997.
Boeing Environmental Technotes. 12: (1). http://www.boeingsuppliers.com/environmental/TechNotes/TNdec07.pdf.
Connor ML. Characterization of recycled carbon fibers and their formation of composites using injection molding. [Master thesis] Raleigh, NC: North Caroline State University; 2008.
Elghazzaoui H. Contribution à l'étude de la dégradation des composites carbone/époxy par solvolyse dans l'eau subcritique et supercritique en vue de leur recyclage[PhD thesis]. France: Université de Nantes; 2012.

Feraboli P, Norris C, McLarty D. Design and certification of a composite thin-walled structure for energy absorption. Int J Veh Des 2007;44(3/4):247–67.

Feraboli P, Kawakami H, Wade B, Gasco F, DeOto L, Masini A. Recyclability and reutilization of carbon fiber fabric/epoxy composites. J Compos Mater 2012;46:1459–73.

George PE. Proceedings of the carbon fibre recycling and reuse 2009 conference. Hamburg, Germany: IntertechPira; 2009.

Heil JP, Cuomo JJ. Study and analysis of carbon fiber recycling. [Master thesis]Raleigh, NC: North Carolina State University; 2011.

Heil JP, Hall MJ, Litzenberger DR, Clearfield R, Cuomo JJ, George PE, Carberry WL. Proceedings of the 2009 SAMPE conference. Baltimore, MD: SAMPE; 2009.

Heil JP, Gavin JB, George PE, Cuomo JJ. Proceedings of the 2011 SAMPE conference, Long Beach, CA, 23–26th May 2011, paper no.1249. Covina, CA: SAMPE International Business Office; 201115.

Hofmann M, Schreiber J. Processing of recycled carbon fibers into nonwovens. Tech Text 2012;55(2):66.

Howarth J, Jeschke M. Proceedings of the carbon fibre recycling and reuse 2009 conference. Hamburg, Germany: IntertechPira; 2009.

Illing-Günther H, Hofman M, Gulich B. Nonwovens made of recycled carbon fibres as basic material for composites, In: Proceedings of the 7th international CFK-Valley Stade convention "Latest Innovations in CFRP Technology". 2013. http://www.cfkconvention.com/fileadmin/Convention_2013/Referenten/Vortraege/CFK_Conv2013_ILLING-GUENTHER.pdf.

Iwaya T, Tokuno S, Sasaki M, Goto M, Shibata K. Recycling of fiber reinforced plastics using depolymerization by solvothermal reaction with catalyst. J Mater Sci 2008;43:2452–6.

Janney M, Baitche N. Composites and polycon 2007. Tampa, FL: American Composites Manufacturers Association; 2007.

Janney MA, Newell WL, Geiger E, Baitcher N, Gunder T. Proceedings of the 2009 SAMPE conference, 18–21st May. Baltimore, MD: SAMPE; 2009.

Jiang G, Pickering SJ, Walker GS, Wong KH, Rudd CD. Proceedings of the 38th international SAMPE technical conference, Dallas, TX, USA.

Jiang G, Wong KH, Pickering SJ, Walker GS, Rudd CD. Alignment of recycled carbon fibre and its application as a reinforcement, In: Proceedings of the SAMPE fall technical conference: global advances in materials and process manufacturing of composites from recycled carbon fibre engineering, 6–9th November 2006vol. 38. Dallas, TX: International Business Office; 2006b. p. 9 Paper No. 4.

Kouparitsas CE, Kartali CN, Varelidis PC, Tsenoglou CJ, Papaspyrides CD. Recycling of the fibrous fraction of reinforced thermoset composites. Polym Compos 2002;23:682–9.

Liu Y, Liu J, Jiang Z, Tang T. Chemical recycling of carbon fibre. Polym Degrad Stab 2012;97(3):214.

Longana ML, Yu H, Hamerton I, Potter KD. Development and application of a quality control and property assurance methodology for reclaimed carbon fibers based on the HiPerDiF (High Performance Discontinuous Fibre) method and interlaminated hybrid specimens. Adv Manuf Polym Compos Sci 2018;4:48–55.

McConnell VP. Launching the carbon fibre recycling industry, Reinf Plast 2010;(March/April). http://www.reinforcedplastics.com/view/8116/launching-the-carbon-fibre-recycling-industry/.

Meredith J. Proceedings of the carbon fibre recycling and reuse 2009 conference. Hamburg, Germany: IntertechPira; 2009.

Meredith J, Cozien-Cazuc S, Collings E, Carter S, Alsop S, Lever J, Coles SR, Wood BM, Kirwan K. Recycled carbon fibre for high performance energy absorption. Compos Sci Technol 2012;72(6):688–95.

Meyer LO, Schulte K, Grove-Nielsen E. CFRP-recycling following a pyrolysis route: process optimization and potentials. J Compos Mater 2009;43(9):1121.

Molnar A. Recycling advanced composites. Final report for the Clean Washington Center (CWC).

Ogi K, Nishikawa T, Okano Y, Taketa I. Mechanical properties of ABS resin reinforced with recycled CFRP. Adv Compos Mater 2007;16:181–94.

Palmer J, Savage L, Ghita OR, Evans KE. Sheet moulding compound (SMC) from carbon fibre recyclate. Compos A: Appl Sci Manuf 2010;41:1232–7.

Pickering SJ. Recycling technologies for thermoset composite materials—current status. Compos A: Appl Sci Manuf 2006;37:1206–15.

Pickering SJ. Proceedings of the carbon fibre recycling and reuse 2009 conference. Hamburg, Germany: IntertechPira; 2009.

Pickering SJ, Robinson P, Pimenta S, Pinho ST. Proceedings of the meeting of the increasing sustainability and recycling consortium, 10th June 2010. London: BIS—UK composites strategy; 2010.

Pimenta S. Toughness and strength of recycled composites and their virgin precursors[PhD thesis]. London: Imperial College London; 2013.

Pimenta S, Pinho ST. Recycling carbon fibre reinforced polymers for structural applications: technology review and market outlook. Waste Manag 2011;31:378–92.

Pimenta S, Pinho ST, Robinson GS. Proceedings of the 17th international conference on composite materials, July, Edinburgh, UK, 2009.

Pimenta S, Pinho ST, Robinson P, Wong KH, Pickering SJ. Mechanical analysis and toughening mechanisms of a multiphase recycled CFRP. Combust Sci Technol 2010;70:1713–25.

Stoeffler K, Andjelic S, Legros N, Roberge J, Schougaard SB. Polyethylene sulphide (PPS) composites reinforced with recycled carbon fiber. Combust Sci Technol 2013;84:65–71.

Szpieg M. Development and characteristics of a fully recycled CF/PP composite[PhD thesis]. Sweden: Luleå University of Technology; 2011.

Szpieg M, Giannadakis K, Asp L. Viscoelastic and viscoplastic behavior of a fully recycled carbon fibre-reinforced maleic anhydride grafted polypropylene modified polypropylene composite. J Compos Mater 2012;46(13):1633–46.

Takahashi J, Matsutsuka N, Okazumi T, Uzawa K, Ohsawa I, Yamaguchi K. Mechanical properties of recycled CFRP by injection molding method, In: Proceedings of the 16th international conference on composite materials 8–13 July 2007, Kyoto, Japan; 2007.

Thompson MR, Tzoganakis C, Rempel GL. Terminal functionalization of polypropylene via the Alder Ene reaction. Polymer 1998;39(2):327–34.

Thryft AR. Recycled carbon fibre save money materials and assembly, http://www.designnews.com/document.asp?doc; 2010.

Turner TA, Pickering SJ, Warrior NA. Proceedings of the 2009 SAMPE conference, 18–21st May. Baltimore, MD: SAMPE; 2009.

Warrior NA, Turner TA, Pickering SJ. Proceedings of the carbon fibre recycling and reuse, 2009. Hamburg, Germany: IntertechPira; 2009.

Wong KH, Pickering SJ, Brooks R. Proceedings of the composites innovation 2007—Improved sustainability and environmental performance, Organised by NetComposites, 4–5th October, Barcelona, Spain.

Wong KH, Pickering SJ, Turner TA, Warrior NA. Proceedings of the 2009 SAMPE conference. Baltimore, MD: SAMPE; 2009.

Wong KH, Pickering SJ, Rudd CD. Recycled carbon fibre reinforced polymer composite for electromagnetic interference shielding. Compos A: Appl Sci Manuf 2010;41:693–702.

Wong KH, Mohammed DS, Pickering SJ, Brooks R. Effect of coupling agents on reinforcing potential of recycled carbon fibre for polypropylene composite. Combust Sci Technol 2012;72:835–44.

Yip HLH, Pickering SJ, Rudd CD. Characterisation of carbon fibres recycled from scrap composites using fluidised bed process. Plast Rubber Compos 2002;31(6):278–82.

Yuyan L, Guohua S, Linghui M. Recycling of carbon fibre reinforced composites using water in subcritical conditions. Mater Sci Eng A 2009;520:179–83.

Further reading

Janicki Industries Inc. www.janicki.com; 2010.

Janney M, Geiger Jr. E, Baitcher N. Fabrication of chopped fiber preforms by the 3-DEP process. In: Proceedings of the composites & polycon conference, 17–19 October 2007, Tampa, Florida, USA. 2007.

Marsh G. Reclaiming value from post-use carbon composite. Reinf Plast 2009;53(4):22.

Materials Innovation Technologies LLC. Fletcher, NC, USA. www.nemergingmit.com.

Sigmatex Ltd. http://www.sigmatex.com/Solution/Recycled_Fabrics.

Technical Fibre Products (TFP). http://www.tfpglobal.com/materials/carbon/.

Technical Fibre Products Ltd. www.techfibres.com; 2010.

CHAPTER 8

Applications of carbon fiber/carbon fiber-reinforced plastic/recycled carbon fiber-reinforced polymers

Carbon fibers are the predominant high strength, high modulus reinforcing fibers used for manufacturing advanced composite materials. Carbon fibers are available in a variety of useful forms. These are thin filaments, and the thickness is one-tenth of a human hair. The fibers are bundled, woven, and shaped into tubes and sheets (up to half inch thick) for certain applications, supplied as cloth for molding, or just conventional thread for filament winding.

High-performance carbon fiber composites need excellent mechanical properties for their application in aerospace and aeronautics, transportation, sport and recreation, compressed gas storage, and civil engineering. Properties of a carbon fiber composite are dependent upon the properties of its constituents and also depend on the properties of the interface/interphase (Ozcan et al., 2014). Depending on the nature of the matrix, carbon fiber composites are used in a variety of applications (Table 8.1). These are used as a structural reinforcement in composites with a broad range of mechanical properties and hence value. Carbon fibers are used where low weight, high stiffness, high conductivity, or where the appearance of the carbon fiber weave is most wanted. Carbon fibers are becoming part of our everyday lives (Fitzer, 1990; Chung, 1994; Watt, 1985; Patel, 2010).

> *Carbon fiber reinforced (CFC) are competing and becoming cost effective in comparison to metals. At the current prices, CFC components are expensive in comparison with metal components. Many developments in raw materials, manufacturing technologies, assembly methods are affecting directly the cost of composites design and development. These advanced techniques will help to reduce the cost of composites significantly which will simulate the requirement for composites exponentially in coming years. Composites design, analysis, manufacturing tools will help in reducing the engineering cycle time, reduce the costs and improve the quality while maintaining repeatability of parts being manufactured.*
>
> **(Shama Rao et al., 2017; www.infosys.com)**

Carbon fiber industry has been growing gradually for meeting the demand from industries such as aerospace (aircraft and space systems), wind, turbine blades, automotive,

Table 8.1 Current applications of carbon fiber composites.

Aerospace; space, military, and commercial
Industrial use
Sports and leisure
Energy wind blades
Energy storage, flywheels
Medical implants (prostheses), X-ray, and MRI equipment

military, marine, construction (nonstructural and structural systems), lightweight cylinders and pressure vessels, offshore tethers and drilling risers, medical, automobile, and sporting goods (Soutis, 2005; Ogawa, 2000; Roberts, 2006; Red, 2006; Pimenta and Pinho, 2011; Carson, 2012; Hajduk and Lemire, 2005). Carbon fiber composites are also used in rockets and satellites. In athletics, they are used in fishing rods, golf club shafts, and tennis rackets for achieving reduction in weight or enhancement of rigidity and strength. In medical application, by utilizing its radiolucent features, CFRP that are lightweight and easily handled are used for medical devices such as limb prostheses and wheelchairs. These are frequently used for X-ray devices (Saito et al., 2011). Moreover, carbon fiber finds use in construction and civil engineering in materials for seismic strengthening or construction.

8.1 Applications

8.1.1 Wind turbine blades

In most of the wind turbine blades, fiberglass is used. On large blades (often over 150 ft in length) include a spare, which is a stiffening rib that runs along the length of the blade. These components mostly contain 100% carbon and as thick as a few inches at the root of the blade. Carbon fiber provides the required rigidity, without adding a significant amount of weight. This is quite important as the lighter wind turbine blade is more capable in generating electricity (http://www.thoughtco.com).

> *The manufacture of wind turbine blades is forecast to be the fastest growing single application for carbon fibre usage in the industrial market sector over the next few years.*
> **(www.plasticbiz360.com)**

Vestas manufacturer of wind turbine is using CFRPs for wind turbine blades. Actually, reduced weight allows for larger blades that is perfect for offshore sites and more suitable in lighter wind conditions.

8.1.2 Automotive

Automobiles produced on a mass scale are not yet using carbon fiber. This is due to the high cost of raw material and essential changes in tooling, still, outweighs the benefits.

But Formula 1, NASCAR, and high-end cars are using carbon fiber. In several cases, it is not due to the advantages of properties or weight, but due to the appearance. Several aftermarket automotive parts are being produced using carbon fiber, and they are clear coated instead of being painted. The clearly different carbon fiber weave has become a sign of high tech and high performance. Actually, it is common to see an aftermarket automotive component that is a single layer of carbon fiber but has several layers of fiberglass below for reducing the cost. This would be an example where the appearance of the carbon fiber is essentially the deciding factor (Brosius, 2003; www.thoughtco.com).

Carbon fiber/epoxy composites have proved in difficult service applications such as military helicopters and jet fighters for over two decades. The material possess infinite fatigue strength, providing strain, is kept to a reasonable level (0.3%). The glass transition temperatures of epoxy resins used approaches and in some cases exceeds 150°C. This makes them suitable for use in engine compartments and primary vehicle structure. In crash situations, carbon fiber has allowed many race car driver to survive, certainly, in most cases, the ability to walk away from a high-speed accident. Carbon fiber components used in production vehicles have passed a number of tests such as rough road durability, crash, simulated hail testing, and hot/cold slamming (www.speautomotive.com).

Europe is leading the way in using carbon fiber for automobiles. In Europe, several vehicles are in production today. The most noticeable of these are the 200 mph "super cars," including the Lamborghini Murciélago, Porsche Carrera GT, and McLaren Mercedes SLR, all scheduled for production volumes of 500 per year. Taking into consideration that roughly every body panel and, in the cases of the Carrera GT and SLR, the bulk of the structure is carbon fiber, and the volume consumed is quite attractive to material suppliers. In the United Kingdom, several niche producers, including TVR, Invicta, Farboud, and MG Rover, are taking benefit of the low capital costs in tooling for carbon fiber composites. Each is also using new materials developed for vacuum bag cure only, avoiding the requirement for expensive autoclaves (www.speautomotive.com).

> *In the current market for carbon fiber automobiles, preimpregnated tapes and fabrics (prepregs) are the dominant form of starting material. A primary reason for this is these materials have become the standard for building open wheel racing vehicles, such as Formula 1, IndyCar and CART, and the design techniques for these vehicles can also be applied to high performance sports cars. Prepregs have inherently higher fiber volumes (typically 50 to 60 percent) than wet layup methods, resulting in higher stiffness at lower weight. Prepreg technology was developed initially for the aerospace industry, and the materials offer very tight control of resin content and fiber areal weight, resulting in consistent, predictable properties and cured ply thickness.*
>
> **(www.speautomotive.com)**

The exploitation of CFC in the automotive industry is ever-increasing as new materials and new manufacturing processes are becoming available (Donnet and Bansal, 1990; Minus and Kumar, 2005, 2007; Fitzer et al., 1989).

> *Optimistic levels of growth are also predicted for the automotive, industrial and power generation sectors. Current levels of carbon fibre usage are in excess of 100,000 ton per annum with growth forecast to be between 10% and 20% per annum. Automotive remains an interesting potential future user of considerable amounts of fibre with global production now in excess of 90,000,000 vehicles per annum. Even a small amount of fibre in a small fraction of total production would make a significant impact on demand.*
>
> **(Pickering et al., 2016)**

Several companies have developed commercial products. "Quantum Composites (Bay City, Michigan, Unied States), a subsidiary of The Composites Group, has launched a hybrid carbon-glass fiber composite material. This is believed to be a high-strength, cost effective and lightweight substitute for conventional metal and glass fibre applications in the automotive industry This brand named hybrid material, AMC-8590-12CFH, is suitable for fast-cure compression molding for manufacturing complex parts on large scale" (http://www.compositesworld.com/products/hybrid-carbon-fiberglass-fiber-reinforcement; http://www.quantumcomposites.com).

Department of Energy, the United States, in collaboration with several universities and research institutions has started the Institute for Advanced Composites Manufacturing Innovation (IACMI). The founding research partners were The University of Tennessee, Knoxville, and Oak Ridge National Laboratory. "Funded through $70 million in federal and $180 million in non-federal funds, IACMI will focus on increasing production capacity of carbon fibre and developing less expensive but advanced fiber-reinforced polymer composites for automotive and other industrial sectors. Incorporating short and nano fibres can certainly exploit the fullest potential of high strength carbon material in hybrid design as a measure to have better properties and economize the formulation" (Sloan, 2015).

> *In 2020, CFRPs across all applications will comprise a $35 billion market, including $6 billion in automotive CFRPs, according to Lux Research. However, that automotive use will be limited to luxury and racing vehicles. Analyses indicate that large-scale, mainstream CFRP automotive adoption before 2020 is unlikely. But sometime after 2020, the potential volume of CRFP used in cars and trucks could dwarf all other applications, potentially reaching hundreds of billions of dollars.*
>
> **(machinedesign.com)**

> *The demand for CFRPs is growing apace, with the automotive sector, for example, predicted to almost triple its usage from $ 2.4 billion in 2015 to $ 6.3 billion in 2021. These needs are currently met by virgin CF (vCF), estimated to have grown from around 16 000 ton a year to 140 000 ton by 2020. But the high financial and environmental cost of vCF is still limiting greater mass-market adoption.*
>
> **(Meng et al., 2017a,b; www.materialstoday.com)**

Over the last two decades, carbon fiber has found its way into almost every hypercar, supercar, and luxury sports car in production. CFRP is widely used in the high end of the market—motorsport, supercar, and premium sports cars:

- In these segments of the market, car performance is the critical factor; weight reduction, lower CoG, driver experience, and aesthetics are the main drivers.
- Carbon fiber is in commonplace use in components including primary structure (monocoque), roof structures, and internal and external body panels.
- The majority share of materials is supplied in prepreg state, but with some cases HP-RTM is used (typically in relatively high-volume programs).
- Until recently, there has been little innovation in this side of the market with customers using what they know and trust (http://feiplar.com.br/pos_feira/apresentacoes/dia06/congresso_sampe/congresso_sampe_sigmatex.pdf).

There is also scope for manufacturing automotive components with RCFRP not only for technical or economic reasons but also to enhance environmental credentials. As legislation tightened up regarding recyclability and sustainability (EU 2000/53/EC), the automotive industry became interested in natural composites (Ellison and McNaught, 2000), which are used extensively in mass production in spite of the associated problems (e.g., consistency of feedstock); RCFRP could become an environmentally friendly material with improved mechanical performance.

Structural demonstrators produced with recycled carbon fiber-reinforced polymers (RCFRP) are aimed at aircraft or automotive industries. Other markets that have been identified are the construction industry, sports and household goods, and wind turbines (Roberts, 2011; Pickering et al., 2010).

The use of composite materials in different parts of aircraft started in the 1970s and made possible the weight reduction of aircraft (Hajduk, 2005; Xiaosong, 2009). Presently, in some types of aircraft, >10% of the total weight is being used, and such applications will perhaps increase in the coming year. The use of carbon fiber-based composites for structural parts in the transportation industry, both car and commercial transportation alike, would increase substantially in the foreseeable future. This is enabled by European legislation centered on carbon dioxide emissions and fuel efficiency, which producers can obtain by reducing the weight of the vehicles and increasing the liking of customers for more ecological vehicles.

In the automobile industry, development of lightweight and fuel efficient cars using CFRP in the structural parts is progressing rapidly. Applications of carbon fiber have also been studied in the energy fields, including the development of fuel cells and oil drilling, as well as in electronic devices, for example, personal computers and liquid-crystal projectors.

8.1.3 Aerospace

Aerospace and space were a few of the first industries to use carbon fiber. It possess high modulus that making it structurally suitable for replacing alloys like aluminum and titanium. Carbon fiber provides weight savings that is the main motive for using carbon fiber in the aerospace industry. Every pound of weight savings can make a major difference in fuel consumption. This is the main reason why new 787 Dreamliner of Boeing is the best-selling passenger plane in history. Bulk of this plane's structure consists of carbon fiber-reinforced composites.

> *Carbon fibre is finding increasing numbers of applications in lightweight structures due to its high specific mechanical properties. High growth rates are already being observed in several major civil aerospace programmes, such as Boeing 787 Dreamliner and Airbus A350.*

Carbon fiber-reinforced polymers have superior strength to weight ratio making it a best choice for aerospace applications, as engineers are making more efficient—and hence lighter—aircraft structures. Although the material offers improvements in some major properties, others are always not wanted.

> *The mechanical properties of composites can be tailored and while many think of it as a very stiff material, it can in fact be made to flex by quite a significant amount. This has become vital in the aviation industry, which have designed advanced wing systems that rely on the ability to flex, quite significantly, to gain aerodynamic performance. Indeed Airbus, Boeing and Bombardier have all been – or are in the process of – developing carbon fibre wings, and this has led to tests to verify the required strength and flexibility has been achieved.*
>
> *(www.materialsforengineering.co.uk)*

In the aerospace sector, Boeing's Dreamliner aircraft relies on CFRP components for reducing weight. With lighter components, higher efficiency and lesser fuel consumption during the lifetime of systems can be obtained resulting in reduced carbon emissions.

In the Boeing Dreamliner 787, the carbon fibers made by Toray Industries account for ≈50% of the aircraft weight. The carbon fibers are used in the fuselage and wings. Relative to comparable models the plane uses 20% less fuel. In June 2011 a new carbon fibers production line started operation at the Mitsubishi Rayon facility in Otake, Hiroshima prefecture. It has a capacity of 2700 tpy. The output goes mostly to producers of wind turbines, for which there is rapidly increasing demand. A plant under development by the Toray Group in South Korea started up in January 2013 and produces carbon fibers for export.

8.1.4 Military

> *There are wide range of applications of Carbon Fiber in the military from planes and missiles to protective helmets, providing strength and weight reduction across all military equipment. It takes energy to move weight – whether it is a soldier's personal gear or a field hospital, and weight saved means more weight moved per gallon of gas. A new military application is announced almost every day. Perhaps the latest and most exotic military application is for small flapping wings on miniaturized flying drones, used for surveillance missions. Of course, we don't know about all military applications – some carbon fiber uses will always remain part of 'black ops' - in more ways than one.*
>
> *(www.engineeringcivil.com)*

8.1.5 Sporting goods

Another market segment that is more than willing to pay more for higher performance is recreational sports. Products commonly produced with carbon fiber-reinforced composites are tennis rackets, golf clubs, softball bats, hockey sticks, and archery arrows and bows. Lightweight equipment without sacrificing strength is a noticeable benefit in sports. For instance, with a tennis racket of lighter weight, it is possible to get much faster racket speed and, finally, hit the ball harder and faster. Athletes are continuing to push for benefit in equipment. This is the reason serious bicyclists ride all carbon fiber bikes and use bicycle shoes using carbon fiber (www.thoughtco.com).

> *Application in sports goods ranges from the stiffening of running shoes to ice hockey stick, tennis racquets, and golf clubs. 'Shells' (hulls for rowing) are built from it, and many lives have been saved on motor racing circuits by its strength and damage tolerance in body structures. It is used in crash helmets too, for rock climbers, horse riders, and motorcyclists – in fact in any sport where there is a danger of head injury.*
>
> *(Todd, 2019)*

8.1.6 Construction industry

Carbon fiber composites are also used in consumer sports goods such as golf shaft, tennis racket, hockey stick, bicycle, softball bat, and also in fishing rods. The consumption of carbon fiber composites internationally is expected to grow at 10%–15% (www.plasticbiz360.com).

Carbon fiber is slowly emerging as a popular alternative to steel and aluminum because of its high strength, high stiffness, and low weight. "Having a lightweight construction allows you to be clever with the actual design of the building. Take Apple's new headquarters, which features the world's largest freestanding carbon fibre roof. This has enabled much of the rest of the building to be made of glass with minimal structural requirements to support it" https://www.themanufacturer.com/articles/five-reasons-why-carbon-fib.

Carbon fiber is being increasingly used to replace steel and aluminum in construction as its low weight and high load bearing capacity make it an ideal material for retrofitment projects.

The average longitudinal tensile strength of carbon fiber is 2000 MPa; elastic modulus is 150–200 GPa. On an average, it is ~2.5 times stiffer than aluminum. It can be produced at various densities in limitless shapes and sizes and has a wide range of applications. Carbon fiber is often shaped into tubes, fabric, and cloth and can be custom made into any number of composite parts and pieces. Because of these factors, carbon fiber is emerging as a popular alternative to steel and aluminum in sectors such as construction, aviation, and automobile. The unique properties of carbon fiber make it an ideal material for construction industry. The construction industry is looking for better alternatives to the materials used currently. This has paved the way for carbon fibers that are increasingly used due to the benefits associated with them. It is a good substitute to steel filaments in fiber concrete, which are usually cheap cellulose or polyacrylonitrile-based fibers.

Carbon fiber has low thermal conductivity and provides good cohesion when used with concrete, which can be beneficial for high loaded floors and roads. It is also five times stronger than steel, while weighs just one-third of it and has a much higher load bearing capacity. This helps in easy delivery of carbon fiber wraps at the construction location as well as makes installation simpler, which reduces the overall timeline of a project. Carbon fiber can be used as a replacement for asbestos in fiber cement as it does not create any inhalation issues.

Global carbon fiber share by application in 2017 is presented in Table 8.2 (https://www.aranca.com/knowledge-library/articles/business-research/is-carbon-fiber-the-future-of-construction-industry).

The global demand for carbon fiber in the construction sector in 2017 was 6000 ton and is expected to be 2200 ton in 2020.

Strengthening is among the major uses of carbon fiber in the construction industry. A number of structural engineering applications use carbon fiber-reinforced polymer due to its potential construction-related advantages and cost efficiency. The common applications include strengthening structures made of concrete, steel, timber, masonry, and cast iron; backfitting to increase the load capacity of old structures such as bridges; and enhancing the shear strength and flexural capacity in reinforced concrete structures. Other applications include uses in prestressing materials, as substitute to steel, and for strengthening cast-iron beams.

For prestressed concrete (PC) structures, the repair technology involves the carbon fiber-reinforced polymer (CFRP) laminate bonding in which a thin flexible fiber sheet adheres to the concrete surface with a thermos-set resin. This increases the shear and flexural capacity of beams and slabs as well as confinement in columns. In case of precast concrete, instead of steel mesh reinforcement, a carbon fiber grid is used in panel faces and

Table 8.2 Global carbon fiber share by application in 2017.

Aviation, aerospace, defense, 30%
Automotive, 22%
Wind energy, 13%
Sports and leisure, 12%
Construction, 5%
Others, 18%

https://www.aranca.com/knowledge-library/articles/business-research/is-carbon-fiber-the-future-of-construction-industry.

as a mechanical link to the outer and inner sections of the concrete wall. This type of carbon fiber grid strengthening reduces the amount of raw materials used as well as the weight.

One of the major challenges impacting the use of carbon fiber in the construction industry, especially in an emerging economy, is the high cost of the material. The cost of carbon fiber is much higher than that of other products such as steel and concrete reinforcement materials. For example, if the average price of steel reinforcement is around USD7–10 per ft^2, the average cost of carbon fiber installation would vary in the range of USD60–100 per ft^2.

Apart from the high initial cost, lack of awareness regarding the overall cost advantage and preference for cheaper traditional products such as steel and concrete have hampered the usage of carbon fiber in structural reinforcement projects.

Despite the challenges, utilization of carbon fiber in construction sector is expected to increase. Factors such as durability, ease of installation, and decline in cost are expected to drive the demand for carbon fiber in the construction sector in future. Carbon fibers can be installed easily without modifying the architecture of a structure and can be applied to the surface of any shape. This reduces the overall project timeline by almost 50%, depending on the size of the project, which brings down the cost (https://www.aranca.com/knowledge-library/articles/business-research/is-carbon-fiber-the-future-of-construction-industry).

Decrease in cost is expected to boost demand for carbon fibers in the construction sector. In the last decade the cost of carbon fiber declined from almost USD150 a pound to USD10 a pound. According to some estimates the cost of carbon fiber may decline by almost 90% in the near future, mainly due to advancement in production techniques and reduction in prices of raw materials. Increasing affordability, coupled with the benefits, is likely to result in the emergence of carbon fiber as a strong replacement for other metals.

8.1.7 Marine Industry

Carbon composites are being considered in selected areas of marine industry for achieving high performance with least weight. Carbon fibers offer vessel stability advantages and high performance with minimum weight. So, these are widely used for sailboats, furniture on superyachts, and high strength interior moldings. Carbon fiber is also seen as a stylish material. It is regularly substituted for higher material characteristics and also for the aesthetics component of the woven material.

8.1.8 Medical applications

Carbon fiber shows many benefits over other materials in the medical field. It is transparent to X-rays and shows as black on X-ray images. It is used extensively in imaging equipment structures for supporting limbs being X-rayed or treated with radiation.

The use of carbon fiber for strengthening of damaged cruciate ligaments in the knee is being researched, but perhaps the best known medical use is that of prosthetics—artificial limbs.

> *South African athlete Oscar Pistorius brought carbon fiber limbs to prominence when the International Association of Athletics Federations failed to ban him from competing in the Beijing Olympics. His controversial carbon fiber right leg was said to give him an unfair advantage, and there is still a lot of debate about this.*
>
> ***(www.engineeringcivil.com)***

Several new applications are seen almost daily (http://www.materialsforengineering.co.uk/engineering-materials-explore/composite-materials/features/carbon-fibre-replacing-metals-and-polymers-as-material-of-choice-in-medical-applications/160312/). The growth of carbon fiber is fast, and in next 5 years, this list is expected to be much longer.

Since the late 1970s carbon fibers have also been investigated for use as biomaterials. The most important benefit of using carbon fiber as biomaterials is their mechanical properties: light weight, high strength, and flexibility (Saito et al., 2011).

Also, carbon fibers are easy to combine with traditional biomaterials and have several fiber morphologies and high radiolucency. Carbon fiber has the required characteristics of biomaterials, such as good biocompatibility. The latest technological developments have realized nanolevel control of carbon fiber. Carbon fiber having diameter in the nanoscale (carbon nanofibers) significantly increases the functions of traditional biomaterials and make possible the development of new composite materials. Carbon nanofibers are opening possibilities for new applications in regenerative medicine and cancer treatment. The first three-dimensional constructions with carbon nanofibers have been possible. These materials could be used as exceptional scaffolding for the regeneration of bone tissues.

Investigation on applications of carbon fibers to biomaterials was started over 30 years back. Several products have been developed. As the newest technological developments have realized nanolevel control of carbon fibers, applications to biomaterials have also progressed to the age of nanosize. Carbon fibers with diameters in the nanoscale considerably improve the functions of traditional biomaterials and make the development of new composite materials possible. Possibilities for newer applications in regenerative medicine and cancer treatment also exist. The first three-dimensional constructions with carbon nanofibers have been possible. These materials could be used as superb scaffolding for bone tissue regeneration.

Carbon fibers have been examined as possible constituents of medical appliances for structural fixation of bone fragments, bone substitutes, cellular growth supports in tissue engineering, etc. (Debnath et al., 2004; Rajzer et al., 2010). Biocompatibility of carbon fibers has been the subject of several studies. Carbon fibers induce the growth of new tissue (Jenkins et al., 1977); however, there were also concerns regarding biocompatibility of carbon fibers (Pesakova et al., 2000; Jenkins, 1978; Rohe et al., 1986). The diverse views regarding biocompatibility of carbon fibers may be explained by the use of different types of carbon fibers of different properties, resulting from various technological parameters (Pesakova et al., 2000; Blazewicz, 2001; Blazewicz and Paluszkiewicz, 2001).

Cellular response to the fibrous carbon material is dependent upon the degree of crystallinity of the material; hence only selected types of carbon fibers are suitable for tissue treatment (Debnath et al., 2004; Blazewicz, 2001; Czajkowska and Blazewicz, 1997). Carbon fibers have been used in the rebuilding of parallel-fibred tissue such as ligaments and tendons. They stimulate proliferation of ordered collagenous fibrous tissue in the direction of the carbon fiber filaments (Jenkins et al., 1977; Demmer et al., 1991). Several fibrous forms of carbon materials have been examined for repairing of hard and soft injuries (Blazewicz, 2001) such as hernia repair (Morris et al., 1998) to fill the tissue losses of rabbit's cartilages (Kang et al., 1991) and also in treatment of bone defects (Blazewicz, 2001).

Carbon fibers are broadly applicable in composite materials for medicine. The use of carbon fibers as the modifying phase increases the strength of the polymer/carbon composite, and the fibers uncovered after polymer resorption process may serve as the scaffold for regeneration of bone tissue (Chlopek et al., 2007). Composite carbon biomaterials have been also used in orthopedics for manufacturing plates, screws, and elements of external stabilizers for osteosynthesis, hip joint endoprostheses, and also some types of medical equipment (Fujihara et al., 2004; Blazewicz et al., 1997; Scotchford et al., 2003; Bokros, 1977; Brooks et al., 2004). Controlled modification of the surface of carbon biomaterials can change the effect of immunological cell response (Czajkowska and Blazewicz, 1997).

8.1.9 Other applications

Research performed by Chalmers University of Technology, Sweden (Chalmers University of Technology, 2018), shows that carbon fibers can work as battery electrodes, storing energy directly (www.sciencedaily.com/releases/2018/10/181018082702.htm). This provides new possibilities for structural batteries, where the carbon fiber becomes part of the energy system. The use of this type of multifunctional material contributes to substantial reduction in weight in the aircraft and vehicles of the future that is a major challenge for electrification (Fredi et al., 2018).

The ultimate tensile strength of short carbon fiber–polypropylene composites is higher by 20% in comparison with similar glass fiber-filled samples (Fua et al., 2000). Synthesis of nanocarbon fiber and their composites have been found to be very energy intensive (up to 12 times more energy consumption) in comparison with steel. However, life cycle studies have shown net savings in energy because of the use of lightweight body parts used in vehicles (Khanna and Bakshi, 2009).

In hybrid formulations, carbon fiber is selectively added into glass or other reinforcement fibers for enhancing performance along lading paths in automotive applications. This is one option for offsetting high cost of carbon fibers. Nevertheless, a true potential of this otherwise excellent lightweight material can be recognized until cost-effective sustainable manufacturing techniques are not introduced. Research institutes and other companies are presently working on following research themes (Pervaiz et al., 2016; www.scirp.org):
- advanced/alternate production methods of producing low-cost carbon fiber from conventional feedstock, such as polyacrylonitrile precursor,
- exploring low cost and renewable precursors for producing carbon fibers,
- recycling of carbon fiber on a large scale.

Sales of carbon fiber composites are expected to exceed 290,000 ton in 2024, which should be worth $31 billion, according to recent report from Global Market Insights Inc. Strong demand from aerospace and defense industries based on its lightweight and higher strength will drive industry growth in composites through 2024. The aerospace industry should see an additional bump in demand for composites as fuel economy, and lightweight airliners will continue to be in increasing demand.

Composites are being used extensively in end-use industries, but the recycling issues faced by the producers, along with supply chain management, could limit the growth of the carbon fiber composites market. Producers will focus on improving their distribution for meeting the increasing demand.

Composites will also benefit from short processing times and better formability. And extended shelf lives offer bulk buying that further support manufacturers.

For recycled carbon fiber-reinforced polymers (RCFRP), the promising applications are noncritical structural components (Pickering, 2006, 2009).

Although there are nonstructural applications for recycled carbon fiber (RCF), for example, industrial paints, construction materials, electromagnetic shielding, high-performance ceramic brake disks, and fuel cells (Panesar, 2009; Howarth and Jeschke, 2009), structural applications could fully make use of the mechanical performance of the fibers, thereby increasing the final value of recycled products. The aeronautics industry is mainly interested in using RCFRPs in the interiors of aircraft (Carberry, 2008, 2009; Georg, 2009) as long as the materials are traceable and their properties are consistent (which may be obtained if the feedstock is manufacturing waste). Certification of recycled materials might not be viable in the short term, and it is recognized that RCFRPs should be allowed to mature in nonaeronautical applications first (Georg, 2009). Nonetheless the involvement of aircraft manufacturers in CFRP recycling (e.g., in Aircraft Fleet Recycling Association (AFRA, 2006) and Process for Advanced Management of End-of-life Aircraft (PAMELA) projects (www.afraassociation.org; www.pamelalife.com/english/results/PAMELA-Life-project_results-Nov08.pdf)) and their efforts in identifying appropriate applications for the recyclates (particularly in aircraft interiors) suggest that RCFRP could be incorporated back into noncritical aeronautical applications in the coming years.

Carbon fibers produced by Toray Group is occupying a high share in the aircraft market (Carson, 2012). Along with aircraft manufacturers the Toray Group has developed carbon fibers and prepregs that satisfy the stringent specifications for aircraft. Toray Group is now the only provider of prepregs to the Boeing Company for the tail wing of its B777 aircraft and the primary structure (the parts that, if broken apart, the plane can no longer fly) of B787 for the main wing, tail wing, and fuselage. Toray Group is also supplying carbon fibers to Airbus SAS, which accounts for 50% of their total use of carbon fiber. For improving the lightening and durability of the aircraft, the used amount of carbon fibers is increasing gradually. The Toray Group will continue to develop technologies proactively as the number one supplier of carbon fibers for aircraft applications. Demand for carbon fibers is increasing rapidly (Anon, 2011).

The 787 Dreamliner from Boeing makes substantial use of carbon fibers because of its strength and lightweight. Environmental applications for carbon fibers include the generation of wind power. The two leading producers of carbon fibers are Toray Industries and Teijin Limited. Since the 1970s these companies have each made investments of JPY100 bn in carbon fibers and are focusing on the automotive sector and low-cost processing methods. Carbon fibers offer 10 times the strength of steel at one-quarter of the weight.

Toray Industries, Honda Motor Company, and Nissan Motor Company are collaborating in developing a new carbon fiber material (Anon, 2008). This group aims to develop a method of mass producing the material that will be used in car bodies. The replacement of most of the steel used in cars with the new material will allow vehicles to be 40% lighter than those available presently. A major impediment to the large-scale use of carbon fibers in cars is its high cost. Carbon fibers cost several thousand JPY per kg, compared with JPY300–400 per kg for aluminum and just over JPY100 per kg for steel. The likelihood of continuing increases in steel prices means that, over time, the difference between the prices of steel and carbon fibers is likely to decrease. There is also further potential for increases in the production of carbon fibers materials. Currently an average car weighs $\approx$ 1350 kg, with 75% of this weight related to steel. This would produce an anticipated 30% reduction in annual carbon dioxide emissions and better fuel efficiency. Other participants in the joint project are the textile manufacturers Toyobo Company and Mitsubishi Rayon Company, the plastic parts manufacturer Takagi Seiko Corporation, and the University of Tokyo. Over the next 5 years, the Ministry of Economy, Trade and Industry is to provide the project with total financial support of JPY2 bn. The group intends to develop a new carbon fiber-based material by blending special resins. The new material should be processed readily using steel processing equipment, thereby facilitating the welding process. To decrease the total production cost of carbon fibers, the group also wants to develop carbon fiber recycling technology.

> One of the other largest potential growth areas for carbon fibre is in filament wound composites for offshore oil and gas industry. To put this in perspective, the worldwide actual carbon fibre output is predicted that the offshore industry alone could require 50,000 ton by 2020.
>
> **(www.plasticbiz360.com)**

Carbon fiber composites are also applied to rockets and satellites. In athletics, they are used in fishing rods, golf club shafts, and tennis rackets for achieving weight reduction or improvement of rigidity and durability. In medicine, by utilizing its radiolucent features, CFRP are frequently used for X-ray devices (Saito et al., 2011).

8.1.10 Uses of carbon fiber in India

India has significant potential for wind energy generation. It has been estimated that additional wind power capacity of 10,500 MW would be set up using wind energy. Use of composites in wind energy sector accounts for 14% of the total composites, and the use of carbon fiber usage is 300–500 kg per blade depending upon size.

Construction and infrastructure is another growth area with enormous potential. India has about 45,000 bridges needing repair. Carbon fibers can be cost-effectively used for imparting reinforcing strength at damaged portion of the bridge.

Carbon fiber composites are being used in different industrial areas such as top panel for X-ray CT scanner; electrical cable core; robot hand; and pressure vessels including CNG tanks, high-performance shafts, and compressor turbine blades.

Carbon fiber is an excellent material for high-performance requirements in these application sectors with low weight and superior strength and stiffness being among its most important properties. Carbon fiber-reinforced plastics (CFRP) offer performance benefits that are unique (www.plasticbiz360.com).

References

Aircraft Fleet Recycling Association. www.afraassociation.org; 2006.
Anon. Nikkei net interactive. 128th July.
Anon. Nikkei.com. 123rd August.
Blazewicz M. Carbon materials in the treatment of soft and hard tissue injuries. Eur Cell Mater 2001;2:21–9.
Blazewicz M, Paluszkiewicz C. Characterization of biomaterials used for bone regeneration by FTIR spectroscopy. J Mol Struct 2001;563–564:147–52.
Blazewicz S, Chlopek J, Litak A, Wajler C, Staszkow E. Experimental study of mechanical properties of composite carbon screws. Biomaterials 1997;18:437–9.
Bokros JC. Carbon biomedical devices. Carbon 1977;15:355–71.
Brooks RA, Jones E, Storer A, Rushton N. Biological evaluation of carbon-fibre-reinforced polybutylene-trephthalate (CFRPBT) employed in a novel acetabular cup. Biomaterials 2004;25:3429–38.
Brosius D. Carbon fiber: the automotive material of the twenty-first century starts fulfilling the promise, In: 3rd annual society of plastic engineers automotive composites conference and exposition; 2003. p. 1–7.

Carberry W. Boeing AERO. Magazie QRT 2008;6 4.08, 2008.
Carberry W. Carbon fibre recycling and reuse 2009 conference. Hamburg, Germany: IntertechPira; 2009.
Carson EG. The future of carbon fibre to 2017. In: Global market forecast. Smithers Apex; 2012.
Chalmers University of Technology. Carbon fiber can store energy in the body of a vehicle, ScienceDaily; 2018. 18 October 2018. www.sciencedaily.com/releases/2018/10/181018082702.htm.
Chlopek J, Morawska-Chochol A, Paluszkiewicz C. FTIR evaluation of PLGA-carbon fibres composite behaviour under in vivo conditions. J Mol Struct 2007;875:101–7. https://doi.org/10.1016/j.molstruc.2007.04.0212008JMoSt.875.101C.
Chung DL. Carbon fibre composites. Boston: Butterworth-Heinemann; 19943–11.
Czajkowska B, Blazewicz M. Phagocytosis of chemically modified carbon materials. Biomaterials 1997;18:69–74.
Debnath UK, Fairclough JA, Williams RL. Long-term local effects of carbon fibre in the knee. Knee 2004;11:259–64.
Demmer P, Fowler M, Marino AA. Use of carbon fibre in the reconstruction of knee ligaments. Clin Orthop Relat Res 1991;271:225–32.
Donnet JB, Bansal RC. Carbon fibres. 2nd ed. New York: Marcel Dekker; 19901–145.
Ellison GC, McNaught R. Research & development report NF0309. London: Ministry of Agriculture, Fisheries and Food; 2000.
Fitzer E. Figueiredo JL, Bernardo CA, RTK B, Huttinger KJ, editors. Carbon res filaments and composites. Dordrecht: Kluwer Academic; 1990. p. 3–4 43–72, 119–146.
Fitzer E, Edie DD, Johnson DJ. Figueiredo JL, Bernardo CA, Baker RTK, Huttinger KJ, editors. Carbon fibres filaments and composites. 1st ed. New York: Springer; 1989. p. 3–41 43–72, 119–146.
Fredi G, Jeschke S, Boulaoued A, Wallenstein J, Rashidi M, Liu F, Harnden R, Zenkert D, Hagberg J, Lindbergh G, Johansson P, Lorenzo Stievano L, Asp LE. Graphitic microstructure and performance of carbon fibre Li-ion structural battery electrodes. Multifunct Mater 2018;1(1)015003. https://doi.org/10.1088/2399-7532/aab707.
Fua SY, Laukeb B, Maderb E. Tensile properties of short-glass-fiber- and short-carbon-fiber-reinforced polypropylene composites, Compos A: Appl Sci Manuf 2000;31:1117–25. https://doi.org/10.1016/S1359-835X(00)00068-3.
Fujihara K, Huang ZM, Ramakrishna S, Satknanantham K, Hamada H. Feasibility of knitted carbon/PEEK composites for orthopedic bone plates. Biomaterials 2004;25:3877–85.
Georg PE. Carbon fibre recycling and reuse 2009 conference. Hamburg, Germany: IntertechPira; 2009.
Hajduk F. Carbon fibres overview, In: Global outlook for carbon fibres 2005. Intertech conferences. San Diego, CA, 11–13 October 2005; 2005.
Hajduk F, Lemire T. Global outlook for carbon fibre 2005, Intertech conferences, San Diego, CA, USA, 11–13th October, 2005.
Howarth J, Jeschke M. Carbon fibre recycling and reuse 2009 conference. Hamburg, Germany: IntertechPira; 2009.
Jenkins DHR. The repair of cruciate ligaments with flexible carbon fibre. A longer term study of the induction of new ligaments and of the fate of the implanted carbon. J Bone Joint Surg 1978;60B(4):520–2.
Jenkins DHR, Forster IW, McKibbin B, Ralis ZA. Induction of tendon and ligament formation by carbon implantation. J Bone Joint Surg 1977;59B:53–7.
Kang HJ, Han CD, Kang ES, Kim NH, Yang WI. An experimental intraarticular implantation of woven carbon fibre pad into osteochondral defect of the femoral condyle in rabbit. Yonsei Med J 1991; 32(2):108–16.
Khanna V, Bakshi B. Carbon nanofiber polymer composites: evaluation of life cycle energy use. Environ Sci Technol 2009;43:2078–84. https://doi.org/10.1021/es802101x.
Meng F, McKechnie J, Turner T, Wong KH, Pickering SJ. Environmental aspects of use of recycled carbon fiber composites in automotive applications. Environ Sci Technol 2017a;51(21):12727–36. https://doi.org/10.1021/acs.est.7b04069.
Meng F, Mckechnie J, Turner TA, Pickering SJ. Energy and environmental assessment and reuse of fluidised bed recycled carbon fibres. Compos A: Appl Sci Manuf 2017b;2017(100):206–14.
Minus ML, Kumar S. The processing, properties, and structure of carbon fibers. JOM 2005;57(2):52–8.

Minus ML, Kumar S. Carbon fibre. Kirk-Othmer Encycl Chem Technol 2007;26:729–7495.
Morris DM, Hindman J, Marino AA. Repair of fascial defects in dogs using carbon fibres. J Surg Res 1998;80:300–3.
Ogawa H. Architectural application of carbon fibres: development of new carbon fibre reinforced glulam. Carbon 2000;38:211–26.
Ozcan S, Vautard F, Naskar AK. Designing the structure of carbon fibers for optimal mechanical properties. In: Naskar AK, Hoffman WP, editors. Polymer precursor-derived carbon. Washington, DC: ACS Symposium Series; 2014. p. 215–32.
Panesar S. JEC composites show. Paris, France: JEC Composites; 2009.
Patel MR. Trends carbon fibre—an introduction, magazines. Trends 2010; PlasticBiz360.htm.
Pervaiz M, Panthapulakkal S, Birat KC, Sain M, Tjong J. Emerging trends in automotive lightweighting through novel composite materials. Mater Sci Appl 2016;7:26–38.
Pesakova V, Klezl Z, Balik K, Adam M. Biomechanical and biological properties of the implant material carbon-carbon composite covered with pyrolytic carbon. J Mater Sci Mater Med 2000;11:793–8.
Pickering SJ. Recycling technologies for thermoset composite materials—current status. Compos A: Appl Sci Manuf 2006;37:1206–15. https://doi.org/10.1016/j.compositesa.2005.05.030.
Pickering SJ. Proceedings of the carbon fibre recycling and reuse 2009 conference. Hamburg, Germany: IntertechPira; 2009.
Pickering SJ, Robinson P, Pimenta S, Pinho ST. Meeting of the increasing sustainability and recycling consortium, BIS—UK Composites Strategy, London, UK, 10th June, 2010.
Pickering SJ, Liu Z, Turner TA, Wong KH. Applications for carbon fibre recovered from composites. IOP Conf Ser Mater Sci Eng 2016;139:1–18.
Pimenta S, Pinho ST. Recycling carbon fibre reinforced polymers for structural applications: technology review and market outlook. Waste Manag 2011;31:378–92.
Rajzer I, Menaszek E, Bacakova L, Rom M, Blazewicz M. In vitro and in vivo studies on biocompatibility of carbon fibres. J Mater Sci Mater Med 2010;21:261.
Red C. Aerospace will continue to lead advanced composites market in 2006. Compos Manuf 2006;7:24–33.
Roberts T. The carbon fibre industry: Global strategic market evaluation 2006–2010. Watford, UK: Materials Technology Publications; 200610 93–177, 237.
Roberts T. The carbon fibre industry worldwide 2011-2020: An evaluation of current markets and future supply and demand. UK: Materials Technology Publications; 2011.
Rohe K, Braun A, Cotta H. Ligament replacement with carbon fibre implants in rabbits light- and transmission electron microscopic studies. Z Orthop Ihre Grenzgeb 1986;124:569–77.
Saito N, Aoki K, Usui Y, Shimizu M, Hara K, Narita N, Ogihara N, Nakamura K, Ishigaki N, Kato H, Haniu H, Taruta S, Kim YA, Endo M. Application of carbon fibres to biomaterials: a new era of nano-level control of carbon fibres after 30-years of development. Chem Soc Rev 2011;40(7):3824–34.
Scotchford CA, Garle MJ, Batchelor J, Bradley J, Grant DM. Use of novel carbon fibre composite material for the femoral stem component of a THR system: in vitro biological assessment. Biomaterials 2003;24:4871–9.
Shama Rao N, Simha TGA, Rao KP, Ravi Kumar GVV. Carbon composites are becoming competitive and cost effective. Infosys Limited; 2017.
Sloan J. IACMI consortium formally launched in Tennessee, Industry News; Composites World 2015; Posted 22 June, 2015. http://www.compositesworld.com/news/iacmi-consortium-formally-launched-in-tennessee.
Soutis C. Fibre reinforced composites in aircraft construction. Prog Aerosp Sci 2005;41:143–51.
Todd J. What products use carbon fiber today? ThoughtCo; 2019. 23 February 2019. thoughtco.com/applications-of-carbon-fiber-820384.
Watt W. Kelly A, Rabotnov YN, editors. Handbook of composites—Volume I. Holland: Elsevier Science; 1985. p. 327–87.
Xiaosong H. Fabrication and properties of carbon fibers. Materials 2009;2:2369–403.

Relevant websites

http://feiplar.com.br/pos_feira/apresentacoes/dia06/congresso_sampe/congresso_sampe_sigmatex.pdf.
http://www.compositesworld.com/products/hybrid-carbon-fiberglass-fiber-reinforcement; http://www.quantumcomposites.comwww.thoughtco.com.
http://www.materialsforengineering.co.uk/engineering-materials-explore/composite-materials/features/carbon-fibre-replacing-metals-and-polymers-as-material-of-choice-in-medical-applications/160312/.
https://www.aranca.com/knowledge-library/articles/business-research/is-carbon-fiber-the-future-of-construction-industry.
https://www.themanufacturer.com/articles/five-reasons-why-carbon-fib.
http://linknovate.com.
http://machinedesign.com.
http://www.afraassociation.org.
http://www.ijert.org.
http://www.infosys.com.
http://www.materialsforengineering.co.uk.
http://www.materialstoday.
http://www.materialstoday.com.
http://www.ncbi.nlm.nih.gov.
http://www.pamelalife.com/english/results/PAMELA-Life-project_results-Nov08.pdf.
http://www.plasticbiz360.com.
http://www.sciencedaily.com.
http://www.sciencedaily.com/releases/2018/10/181018082702.htmwww.engineeringcivil.com.
http://www.scirp.org.
http://www.speautomotive.com.
http://www.thoughtco.com.
http://www.toray.com.

Further reading

Composites World. Hybrid carbon fiber/glass fiber reinforcement, Posted on 17 Feb. http://www.compositesworld.com/products/hybrid-carbon-fiberglass-fiber-reinforcement; 2014.
PAMELA. PAMELA-life: Main results of the project. Airbus S.A.S.; 2008.
Quantum Composites. Quantum composites launches first hybrid carbon fiber material, http://www.quantumcomposites.com/; 2014.

CHAPTER 9

The carbon fiber/carbon fiber-reinforced plastic/recycled carbon fiber-reinforced polymer market

"Light in weight, Strong and Durable!" Carbon fibers are a 21st century high-technology material. The fibers have low specific gravity, superb mechanical properties (high specific tensile strength and elastic modulus) and striking performances (electric conductivity, heat resistance, lower thermal expansion coefficient, chemical stability, self-lubrication property, higher heat conductivity, etc.). These features are motivating carbon fiber users for developing various applications (www.formula1-dictionary.net).

> Carbon fibre is the most preferred lightweight manufacturing material and is increasingly becoming the material of choice for the producers around the world. Its high tensile strength, low weight and low thermal expansion have opened a world of opportunities. Everyone from elite athletes, to car enthusiasts, to the makers of passenger jets, are moving to Carbon fibre.
>
> (Fitzer, 1990; Chung, 1994; Watt, 1985; Donnet and Bansal, 1990; Minus and Kumar, 2005, 2007; Fitzer et al., 1989; Hajduk, 2005; Xiaosong, 2009; Roberts, 2006; Carson, 2012; Red, 2006; Pimenta and Pinho, 2011; Anon, 2009).

A map of carbon fiber manufacturing facilities is shown in Fig. 9.1 (Das et al., 2016; Milbrandt and Booth, 2016).

The United States, Western Europe, China, and Japan are the main manufacturing locations. Precursors, specifically polyacrylonitrile, account for more than 50% of the manufacturing cost of carbon fiber, and their high cost is one of the obstacles to their extensive use (Warren, 2014; Rocky Mountain Institute, 2015; DOE (US Department of Energy), 2013; www.nrel.gov). Table 9.1 presents the manufacturing cost breakdown of carbon fiber.

> Carbon fibre is a highly complex industry that produces a wide range of products. In addition to the tow counts, the products have varied properties, and a wide range of resin formulations and combinations are made in the prepreg sector. The consumption of carbon fibre is fragmented across multiple end use applications. It is driven by a combination of drivers: for simplicity these are broken down into aerospace and defence, sports and leisure goods, and industrial applications. The key driver behind all markets is the potential weight savings of using carbon fibres, followed by its performance.
>
> (https://www.smithersapex.com/news/2016/november/carbon-fibre-market-forecast-to-grow-to-132,000-to)

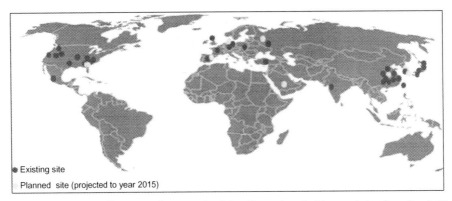

Fig. 9.1 A map of carbon fiber manufacturing facilities. *(Reproduced with permission from Das S, Warren J, West D, Schexnayder S. Global carbon fiber composites supply chain competitiveness analysis. Clean Energy Manufacturing Analysis Center; May 2010. http://www.nrel.gov/docs/fy16osti/66071.pdf.)*

Table 9.1 Carbon fiber manufacturing cost breakdown Rocky Mountain Institute (2015).

Precursor	51%
Utilities	18%
Depreciation	12%
Labor	10%
Other	9%

Based on Rocky Mountain Institute. Carbon fiber cost breakdown; 2015 http://www.rmi.org/RFGraph-carbon_fiber_cost_breakdown.

The application of carbon fiber and carbon fiber-reinforced plastics in aerospace, wind energy, and sports is dominating. The strongest growth rates over the last few years are seen in case of pressure vessels, wind industry, and molding compound sectors. Wind energy and pressure vessels would see the strongest growth over the next few years.

Table 9.2 shows carbon fiber manufacturers, Table 9.3 shows worldwide production capacities of carbon fiber, and Table 9.4 shows production units of different producers in different regions of the world (Gregr, 2010; Roberts, 2006; Carson, 2017).

The major technology developments providing the edge to producers include higher throughput processes and textile polyacrylonitrile precursor use. Usually polyacrylonitrile is a high purity feedstock, and priced at a premium to textile grade. The other development that could facilitate the adoption of carbon fibre reinforced plastics into mass-produced automobiles is technology that allows injection of thermoplastics into part-formed carbon fibre reinforced parts such as organopanels. Thus use of carbon fibre can be synchronised with the automated, high-throughput processes used in the automotive sector, as opposed to causing bottlenecks.

(www.smithersapex.com)

Table 9.2 Manufacturers of carbon fiber.

Toray Industries
Toho Tenax (Teijin)
Mitsubishi Rayon
Mitsubishi Plastics, Inc.
Zoltek
Hexcel
Formosa Plastics
Cytec Engineered Materials
SGL Carbon Group/SGL Technologies
Mitsubishi Chemical
Nippon Graphite Fiber

Based on Gregr J. In: Proceedings of the 7th international conference—TEXSCI 2010, 6–8th September, Liberec, Czech Republic; 2010. Carson, 2017.

Table 9.3 Worldwide production capacities of carbon fibers (metric tons annually).

Manufacturer	Trade name	2010
Small tow PAN-based carbon fibers		
Toray Group	Torayca	20,900
Toho Tenax Co. Ltd.	Tenax	13,500
Mitsubishi Rayon Co., Ltd.	Pyrofil, Grafil	10,100
Formosa Plastics Group	Tairyfil	8750
Hexcel	HexTow	5800
Cytec Engineered Materials	Thornel	3000
Dalian Xingke Carbon Fiber Co., China		760
Aksa Turkey	Aksaca	1500
Large tow PAN-based carbon fiber		
ZOLTEK Group	Panex	13,500
SGL Group	Sigrafil	6000
Pitch-based carbon fibers		
Nippon Graphite Fiber Corporation	Granoc	180
Mitsubishi Plastics, Inc.	Dialead	1250
Cytec Engineered Materials, United States	Thornel	400

Based on Gregr J. In: Proceedings of the 7th international conference—TEXSCI 2010, 6–8th September, Liberec, Czech Republic; 2010.

The carbon fiber/carbon fiber-reinforced plastic/recycled industrial production of carbon fiber began in the 1960s in the United Kingdom, Germany, Japan, and the United States at a total annual production volume of <1000 tons (Warnecke et al., 2010, 2011).

Table 9.4 Production units of different producers in different regions of the world.

Toray	Japan, France, United States
Toho	Japan, BRD, United States
Mitsubishi Rayon	Japan, United States, BRD
Hexcel	United States, Spain
Zoltek	Hungary, United States
SGL	BRD, United States

Based on Gregr J. In: Proceedings of the 7th international conference—TEXSCI 2010, 6–8th September, Liberec, Czech Republic; 2010.

The availability of facilities making polyacrylonitrile precursor fiber with the wet-spinning process enabled increased independence by Japan in the 1980s. The Toray Group and Toho Besion of Japan, which focus on the manufacture of synthetic fibers, have increased carbon fiber production. The two companies, through joint ventures and licenses, have a leading role in global production of carbon fiber. In the future, strong growth in carbon fiber demand is anticipated, but demand in some sectors is slow due to the global economic downturn. Mitsubishi Rayon has, due to this situation, decided to defer for 12 months the start of operations at its new facilities. Expansion of pitch-based carbon fiber capacity is expected, but the overall volume will remain much less than that of PAN precursor-based carbon fiber. Warnecke, (2010) and Anon, (2011) undertook studies of the current situation of carbon fiber producers. They have provided information concerning the eight leading established producers, together with new competitors and those companies planning on entering the market. Only those businesses working with polyacrylonitrile have been considered due to the material dominance for carbon fiber production. Toray Industries (Tokyo, Japan) is the leading producer of carbon fiber worldwide. Initially the company specialized in the production of viscose rayon before starting carbon fiber production in 1971. Toho Tenax Company Limited (Tokyo, Japan) began production of oxidized polyacrylonitrile fiber in 1975 and then started the production of carbon fiber in 1977. Zoltek Companies Incorporated (St. Louis, Missouri, United States) produces carbon fiber and polyacrylonitrile fiber. Mitsubishi Rayon Company Limited (Tokyo, Japan) is Japan's biggest producer of polyacrylonitrile fibers and has been an associated company of the Mitsubishi Chemical Holdings Corporation since 2010. Following these top four producers in terms of production capacities are Formosa Plastics Corporation (Taipei, Taiwan), Hexcel Corporation (Stamford, Connecticut, United States), SGL Group (Wiesbaden, Germany), Cytec Industries (Woodland Park, New Jersey, United States), Dalian Xingke CF (Dalian, China), and Aksa Akrilik Kimya San A (Istanbul, Turkey). The latter is a chemical fiber producer and the world's biggest polyacrylonitrile producer. Other carbon fiber producers have been identified as Dalian Xingke Carbon Fiber, Bluestar Fibers Company Limited China (Beijing, China), Specialty Materials Incorporated (Lowell, Massachusetts, United States), Kemrock Industries and Exports Limited (Vadodara, India), and Fisipe-Fibras Sinteticas de

Portugal, SA (Lavradio, Portugal). With effect from April 2011 shipments, Toray Industries Incorporated has increased the price of carbon fiber by 10%–15% (Anon, 2012). The increases apply to domestic and overseas shipments. Internal prices for general purpose carbon fiber have reached USD28–31 per kg compared with just under USD20 per kg in the first quarter of 2010. Carbon fiber demand in Japan and overseas has increased for use in products such as sporting goods and aircraft.

The benefits of vCF currently come at a price. The production of vCF has a huge impact on the environment; the process is highly energy intensive and mostly relies on fossil fuel-derived precursors. Furthermore, as more CFRP products find a big market, there will be considerably growing requirement for an option to landfill or incineration as components reach the end of their useful life. In the United Kingdom, for instance, these conventional disposal routes presently account for about 98% of composite waste (https://www.materialstoday.com/composite-processing/features/new-lease-of-life-for-cfrps/).

Over the last few years, many changes in the application areas for carbon fiber has been observed. Applications in defense and aerospace applications have grown radically and are now the major consumers of carbon fiber—13,900 tons or 30% based on a total of 46,500 tons. These are followed by products for the wind turbines and sport/leisure sector. Each sector accounts for 14% of total demand (https://www.materialstoday.com/download/79669/). The automotive segment is becoming more and more important. The consumption of carbon fiber has more than doubled over 2013 to about 5000 tons. This is perhaps because of the ramp-up phase for the production of the i-models from BMW. Other applications involve molding and compounding (Mark, 2014).

Because of high strength and high modulus of fibers available, the global carbon fiber market is expected to generate huge revenues in the coming years.

https://qbnnews.com/2019/07/22/carbon-fiber-market-kemrock-industries-hexcel-corporation-formosa-plastic-corporation/.

In developing economies, demand for carbon fibers is increasing. Aside from this the increasing use of carbon in 3D printing would generate new opportunities in the global market. The competitive landscape of the carbon fiber market is literally consolidated because of the presence of very few well-established producers. Carbon fiber is about as half as light as steel and possesses high tensile strength, temperature tolerance, and toughness. It also has low thermal expansion and high heat tolerance and resistance. Due to its distinct lightweight and strength, carbon fibers are an all-round and useful commercial product for a wide range of markets. The carbon fiber market is gaining impetus because of a number of factors like rise in demand for fuel efficient vehicles and launching new products consistently. Moreover the government is laying emphasis to encourage wind energy industry and to increase the usage of materials that are used for manufacturing composite parts of aircraft. Original Equipment Manufacturers (OEMs) have started to substitute traditional materials used for automotive manufacturing with advanced engineered materials like glass fiber composites, engineered plastics, and carbon fiber composites, thus increasing the demand for carbon fibers in the global market.

> *Demand for Carbon fibre will continue to grow, with global demand exceeding 77,000 tons in 2018 and growing to more than 150,000 tons by 2025. Combined global production capacity is currently more than 140,000 tons.*
>
> **(www.futuremarketsinc.com)**

Several producers have increased capacity in 2018, and Kangde (66,000 tons), Toray (62,000 tons), and SGL Group (49,000 tons) will greatly grow their capacity over the next few years. Other companies with expansion plans include Hexcel, DowAska, and Hyosung. Toray purchased carbon fiber component manufacturer TenCate Advanced Composites Holding in March 2018 for $1 billion.

Factors in the growth include increased use in products such as turbine blades for wind farms and high-transmission power lines. Carbon fiber is made from a petrochemical: acrylonitrile. International prices for acrylonitrile have increased by 20%–30% in the past 12 months. Another factor in the price increase of carbon fiber is the reduced export profit margin resulting from a stronger yen. Toray Industries is the leading producer of acrylonite-based carbon fiber. With domestic rivals Toho Tenax Company and Mitsubishi Rayon Company, it accounts for $\approx$70% of the world market.

Toray Industries, Inc. is increasing large tow carbon fiber production capacity at their Zoltek Companies, Inc. based in the United States (https://cs2.toray.co.jp). They have planned to increase production capacity by 50% at Zoltek's Hungarian facility. The total investment exceeds 130 million USD, and production is expected to start in 2020. Zoltek has increased the capacity of their Mexican facility from 5000 tons per year to 10,000 tons per year. Once the production capacity at the Hungarian facility is increased, Zoltek's total annual production capacity will be about 25,000 tons.

Demand for large tow carbon fiber for industrial applications is swiftly increasing. In South America and in Asia, especially China and India, demand for wind energy turbine blades (the major application of Zoltek's large tow carbon fiber) is increasing. Because of the large size of the turbines, the demand for large tow carbon fiber will also increase, resulting in further expansion, particularly in Europe, where carbon fiber is mostly used for automotive structural applications. Zoltek has established a timely supply chain from the Hungarian facility, which is increasing its production capacity. By responding to future increase in demand, Toray Group, including Zoltek, is aiming to become the global standard for automotive applications. Zoltek is becoming a number one provider of large tow carbon fiber in the world and is planning to increase their carbon fiber production equipment at their Hungarian and Mexican facilities.

> *In the medium-term management project—Project AP-G 2019, Toray places the carbon fiber composite material business, a strategically expanding venture, and plans to proceed with business*

expansion by introducing active managerial resources. Toray will focus on expanding in areas that can demonstrate their respective strengths with regular tow2 and large tow carbon fiber and will aim for further expansion of business by developing synergistic effect opportunities that take advantage of the merits of both products.

(https://omnexus.specialchem.com/news/industry-news/toray-increases-large-tow-carbon-fiber-production-capacity-000214244)

9.1 Markets for recycled carbon fiber

More than 30% of the carbon fiber produced eventually ends up as waste, including 22% in the form of manufacturing waste from processing of thermoset carbon fiber-reinforced composites. Carbon fiber demand forecasts to grow to at least 117,000 tons by 2022 according to AVK, and at least 26,000 tons of this carbon fiber would be expected to end up as manufacturing waste according to carbon fiber recycler ELG Carbon Fiber Ltd. (https://www.plasticstoday.com/recycling/recycled-carbon-fiber-key-mass-market-applications/7344798259925).

Reinforced polymer-recycled carbon fibers (RCF) are usually milled or chopped. The milled fiber typically has an average length of 100–150 μm. Chopped fiber can be from 1/8 to 1 in. in average length. Depending on the grade, chopped fiber is usually more valuable than milled fiber due to the length. There are various markets for each of these types of fiber. Although research has shown that RCF retains much of its mechanical properties when compared with virgin fibers, the carbon fiber would not be reused in loadbearing structural applications such as a wing on a Boeing 787 Dreamliner. Milled fiber is used as composite reinforcement for shielding, dissipative, and conductive compounds. The material property of interest is the electrical conductivity of the carbon fiber. If the RCF contains similar conductivity to the virgin carbon fiber, then it can replace the virgin material at a reduced cost. Conductive paper applications are also evaluating the use of recycled carbon fiber and the fuel cell industry for conductive plating due to carbon fiber electrical conductivity. This type of material would be reintroduced into polymers using an extrusion line, thereby forming pelletized material and subsequently injection molded into various parts (Edward, 2008). Chopped carbon fiber can also be utilized in electromagnetic interference shielding applications and also in more traditional components requiring more load bearing. One application, which is growing in use at ≈50% per year, is the buoyant housing material used for oil drilling. Part producers are in requirement of a carbon fiber source for the housing due to the difficulty of obtaining virgin carbon fiber. The major material property of interest here is the mechanical strength of carbon fiber used as reinforcement material. RCF can be used in different market sectors. Depending on the property of interest of the material, RCF can replace virgin carbon fiber to create a cost-effective product. RCF can also be used in applications in which glass fibers are normally used. However, this is dependent upon the price per pound of RCF compared with glass fibers and also the mechanical enhancement needed. If part manufacturers want to replace metals with carbon fiber-reinforced polymers but the price of virgin carbon fiber is too high, they can use RCF at a much lower price, thereby benefiting from the properties of the RCF.

> The promising applications for rCFRPs consists of non-critical structural components Although there are presently non-structural applications for rCFs e.g. industrial paints, construction materials, electromagnetic shielding, high performance ceramic brake discs, fuel cells structural applications would fully utilize the mechanical performance of the fibres, thus increasing the final value of recycled products.
> **(George, 2009; Pickering et al., 2006; Pickering, 2009; Curry, 2010; Howarth and Jeschke, 2009; Panesar, 2009)**

The aeronautics industry is using rCFRPs in the interiors of aircraft as long as the materials are traceable and their properties constant (which may be easily obtained when the feedstock is manufacturing waste) (Carberry, 2009; George, 2009). In the short term, certification of recycled materials might not be viable, and it is recognized that rCFRPs should be allowed to mature in nonaeronautical applications first (George, 2009); nonetheless, the involvement of aircraft manufacturers in CFRP recycling (e.g., in AFRA, 2006 and PAMELA, 2008 projects) and their efforts in finding suitable uses for the recyclates (particularly in aircraft interiors) proposes that rCFRPs might be included back into noncritical aeronautical applications in the near future.

There is a scope of producing automotive components with rCFRPs, for technical or economic reasons, and also to increase green credentials. As there are strict legislation regarding recyclability and sustainability (EU 2000/53/EC 2000), the automotive industry developed interest for natural composites (Ellison and McNaught, 2000), which are these days extensively used in production on a large scale in spite of some associated problems (e.g., consistency of feedstock); rCFRPs could follow as an ecofriendly material showing better mechanical performance.

Presently, structural demonstrators produced with rCFRPs are used in aircraft or automotive industries (Fig. 9.2, Table 9.5). Other areas that have been identified are wind turbines, sports and household goods, and construction industry (Pickering et al., 2006, 2010).

Table 9.6 presents an overview of potential applications for various types of rCFRPs; this is complemented by precise applications presently produced with virgin materials for allowing a direct comparison regarding production methods and mechanical properties.

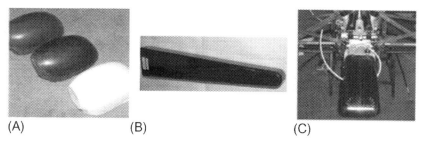

Fig. 9.2 Examples of demonstrators manufactured with recycled CFs: (A) Wing mirror covers (BMC compression, Warrior et al., 2009), (B) Aircraft seat arm-rest (3-DEP process, George, 2009), and (C) Rear or WorldFirst F3 car (woven re-impregnation, Meredith, 2009). *(Reproduced with permission from Pimenta S, Pinho ST. Recycling carbon fibre reinforced polymers for structural applications: technology review and market outlook. Waste Manag 2011;31:378–92.)*

Table 9.5 Demonstrators manufactured with recycled CFs.

rCFRP demonstrator[a]	Virgin component	CF recycler	rCFRP manufacturer	rCFRP matrix[b]
Wing mirror cover	Unknown	Unknown	Warrior et al. (2009)	UP
Car door panel	Unknown	Unknown	Warrior et al. (2009)	EP
Corvette wheelhouse	EoL F-18 aircraft stabilizer	RCFL (2009)	Janney et al. (2009)	UP
Aircraft seat arm rest (George, 2009)	Aircraft testing and manufacturing waste	Janney et al. (2009)	Janney et al. (2009)	EP
Driver's seat of a Student Formula SAE car	Tennis rackets	Nakagawa et al. (2009)	Nakagawa et al. (2009)	UP
Rear structure of WorldFirst F3 *green* car	Outdated woven prepreg	RCFL (2009)	Meredith (2009)	EP
Tool for composite lay-up	Outdated woven prepreg	RCFL (2009)	George (2009)	un.

[a]Additional information can be found in the reference associated with the rCFRP manufacturer, unless otherwise here specified.
[b]Key for rCFRP matrix: *EP*, epoxy resin: *UP*, unsaturated polyester; *un.*, unknown.
Reproduced with permission from Pimenta S, Pinho ST. Recycling carbon fibre reinforced polymers for structural applications: technology review and market outlook. Waste Manag 2011;31:378–92.

Table 9.6 Potential structural applications for rCFRPs.

Type of rCFRP[a]	Possible processes			Examples of current solutions with virgin materials		
	Recycling	Manufacturing	Foreseen markets for rCFRP	Component	Material[b]	ET (GPa)	X_T (MPa)

Type of rCFRP[a]	Recycling	Manufacturing	Foreseen markets for rCFRP	Component	Material[b]	ET (GPa)	X_T (MPa)
Reinforced TP	• Pyrolysis • Fluid. bed • Chemical	• Injection molding • C. mold, non-woven mats	• Automotive semi-structural parts • Construction materials • Leisure and sports goods	• Car dashboard[c] • Car underbody shielding[d] • Plywood replacement[e] • Swimming goggles[f]	GF+PP GF+PP GF+PP GF+PP	5.3 4.5 15	100 60 200
Low-reinforced TS	• Pyrolysis • Fluid. bed • Chemical	• BMC compression • C. mold. Non-woven mats	• Automotive semi-structural parts • Equipment housing	• Car door panel[g] • Carburetor housing[h] • Car headlamp reflectors[i] • Fridge handler[k]	GF+UP GF+UP GF+UP GF+UP	14 13 12[j] 12[j]	36 37 48 44
Medium reinforced TS	• Pyrolysis • Fluid. bed • Chemical	• C. mold. Non-woven mats	• Automotive noncritical structures • Aircraft interiors	• Car rear wing[l] • Car decklid[n] • Aircraft seat structure[o] • Aircraft overhead bin[o]	CF+VE GF+un. Al 2024 GF+PF	27[m] 12 73[p] 20	152[m] 120 469[p] 167
Aligned TS	• Pyrolysis • Chemical	• C. mold. aligned mats • Layup of aligned prepregs	• Automotive noncritical structures • Aircraft interiors • Wind turbine noncritical structures	• Car roof shell[l] • Wind turbine non-critical layers[q]	CF+EP GF+EP	31[m] 22	234[m] 367

Woven TS	• Pyrolysis (pre-preg rolls)	• Resin infusion • RTM	• Automotive structures • Wind-turbine structures	• Car body panels[l] • Wind turbine critical layers[q]	CF+EP GF+EP	53[m] 38	450[m] 711

[a] Key for type of rCFRP: *TP*, thermoplastic matrix; *TS*, thermosetting matrix.
[b] Key for materials: *Al*, aluminum; *CF*, carbon fiber; *EP*, epoxy resin; *GF*, glass fiber; *PA*, polyamide; *PF*, phenolic resin; *PP*, polypropylene; *UP*, unsaturated polyester; *VE*, vinylester resin; *un.*, unknown.
[c] Borealis (2009).
[d] Quadrant (2007).
[e] Quadrant (2010).
[f] Blueseventy (2010).
[g] Menzolit (2004a).
[h] Menzolit (2004b).
[i] BMCI (2010a).
[j] Flexural property.
[k] BMCI (2010b).
[l] Lexus LFA Press Information (2009).
[m] Calculated from specific properties, assuming same density as rCFRPs under comparison.
[n] Menzolit (2008).
[o] Wong et al. (2007).
[p] Aluminum is two times denser than rCFRPs under comparison.
[q] Kong et al. (2005).

Reproduced with permission from Pimenta S, Pinho ST. Recycling carbon fibre reinforced polymers for structural applications: technology review and market outlook. Waste Manag 2011;31:378–92.

References

AFRA. Aircraft Fleet Recycling Association, O_cial website at: www.afraassociation.org; 2006.
Anon. Technical textile 2009; 52, 3, 88.
Anon. Nikkei.com, 2011; 23rd August. 1116.
Anon. Chemical Fibres International, 2012; 62, 2, 61.
Blueseventy. Carbon race goggles, Blueseventy; 2010. Available at: www.blueseventy.com/products/detail/carbon_race.
BMCI. BMC 304 data sheet, Bulk Molding Compounds; 2010a. Available at: www.bulkmolding.com/datasheets/informational/BMC_304.pdf.
BMCI. BMC 310 data sheet, Bulk Molding Compounds; 2010b. Available at: www.bulkmolding.com/datasheets/informational/BMC_310.pdf.
Borealis. Case study: BMW 7 series dashboard carrier, Borealis Group; 2009. Available at: www.borealisgroup.com/pdf/literature/borealis-borouge/case-study/MY_CAST097_GB_2010_02_BB.pdf.
Carberry W. Aerospace's role in the development of the recycled carbon fibre supply chain, In: Carbon fibre recycling and reuse 2009 conferenceHamburg, Germany: IntertechPira; 2009.
Carson EG. The future of carbon fibre to 2017, Global market forecast. Smithers Apex; 2012.
Chung DL. Carbon fibre composites. Boston: Butterworth-Heinemann; 1994, pp. 3–11.
Curry R. Successful research programme allows high performance brake discs to be manufactured at lower cost. Press release log (January), 2010; 1–2.
Das S, Warren J, West D, Schexnayder S. Global carbon fiber composites supply chain competitiveness analysis, Clean Energy Manufacturing Analysis Center; May 2016. http://www.nrel.gov/docs/fy16osti/66071.pdf.
DOE (US Department of Energy). Renewable, low-cost carbon fiber for lightweight vehicles. Summary report from the June 4–5, 2013 workshop, Detroit, Michigan; 2013.
Donnet JB, Bansal RC. Carbon fibres. 2nd ed. New York: Marcel Dekker; 1990, pp. 1–145.
Edward AB. Characterization of reclaimed carbon fibre and their integration into new thermoset polymer matrices via existing composite fabrication techniques, materials science and engineering. [Thesis for masters degree]NC, USA: North Carolina State University; 2008.
Ellison GC, McNaught R. The use of natural fibres in nonwoven structures for applications as automotive component substrates. Research & development report NF0309. Fisheries and Food, UK: Ministry of Agriculture; 2000.
Fitzer E. Figueiredo JL, Bernardo CA, RTK B, Huttinger KJ, editors. Carbon fibres filaments and composites. Dordrecht: Kluwer Academic; 1990. p. 3–4, 43–72, 119–146.
Fitzer E, Edie DD, Johnson DJ. Figueiredo JL, Bernardo CA, Baker RTK, Huttinger KJ, editors. Carbon fibres filaments and composites. 1st ed. New York: Springer; 1989. p. 3–41, 43–72, 119–146.
George PE. End user perspective: perspective on carbon fibre recycling from a major end user, In: Carbon fibre recycling and reuse 2009 conferenceHamburg, Germany: Intertech-Pira; 2009.
Gregr J In: Proceedings of the 7th international conference—TEXSCI 2010, 6–8th September, Liberec, Czech Republic, 2010; 2010.
Hajduk F. Carbon fibres overview. Global outlook for carbon fibres 2005, In: Intertech conferences San Diego, CA, 11–13 October 2005; 2005.
Howarth J, Jeschke M. Advanced non-woven materials from recycled carbon fibre, In: Carbon fibre recycling and reuse 2009 conferenceHamburg, Germany: IntertechPira; 2009.
Janney MA, Newell WL, Geiger E, Baitcher N, Gunder T. Manufacturing complex geometry composites with recycled carbon fiber, In: SAMPE'09 conferenceBaltimore, MD, USA: SAMPE; 2009.
Kong C, Bang J, Sugiyama Y. Structural investigation of composite wind turbine blade considering various load cases and fatigue life. Energy 2005;30:2101–14.
Mark Holmes. Global carbon fibre market remains on upward trend, Reinforced plastics november/december, 2014; www.materialstoday.com.
Lexus LFA Press Information. Toyota (GB) PLC—Lexus Division, Available at: www.toyota.co.jp/en; 2009.

Menzolit. Menzolit BMC 0400 data sheet, Menzolit GmbH; 2004a. Available at: www.menzolit.com/templates/rhuk_solarflare_ii/pdf/list_bmc/BMC_0400.pdf.

Menzolit. Menzolit BMC 1400 data sheet, Menzolit GmbH; 2004b. Available at: www.menzolit.com/templates/rhuk_solarflare_ii/pdf/list_bmc/BMC_1400.pdf.

Menzolit. Menzolit SMC 1800 data sheet, Menzolit GmbH; 2008. Available at: www.menzolit.com/templates/rhuk_solarflare_ii/pdf/list_smc/SMC_1800.pdf.

Meredith J. The role of recycled carbon fibre composites in motorsport applications, In: Carbon fibre recycling and reuse 2009 conferenceHamburg, Germany: IntertechPira; 2009.

Milbrandt A, Booth S. Carbon fiber from biomass. Clean Energy Manufacturing Analysis Center (CEMAC) contract no. DE-A C36-08GO28308 technical report NREL/TP-6A50-66386.

Minus ML, Kumar S. The processing, properties, and structure of carbon fibers. JOM 2005;57(2):52–8.

Minus ML, Kumar S. Carbon fibre. Kirk-Othmer Encycl Chem Technol 2007;26:729–7495.

Nakagawa M, Shibata K, Kuriya H. Characterization of CFRP using re-covered carbon fibers from waste CFRP, In: Second international symposium on fiber recycling, the Fiber Recycling 2009 Organizing Committee, Atlanta, Georgia, USA; 2009.

PAMELA. PAMELA-life: main results of the project, Airbus S.A.S., Available at: www.pamelalife.com/english/results/PAMELA-Life-project_results-Nov08.pdf; 2008.

Panesar S. Converting composite waste into high quality reusable carbon fibre. In: JEC composites show. Paris, France: JEC Composites; 2009.

Pickering SJ. Carbon fibre recycling technologies: what goes in and what comes out?, In: Carbon fibre recycling and reuse 2009 conferenceHamburg, Germany: IntertechPira; 2009.

Pickering SJ, Turner TA, Warrior NA. Moulding compound development using recycled carbon fibres, In: SAMPE fall technical conferenceDallas, USA: SAMPE; 2006.

Pickering SJ, Robinson P, Pimenta S, Pinho ST. Applications for recycled carbon fibre, In: Meeting of the increasing sustainability and recycling consortium (10 June 2010), BISLondon, UK: UK Composites Strategy; 2010.

Pimenta S, Pinho ST. Recycling carbon fibre reinforced polymers for structural applications: technology review and market outlook. Waste Manag 2011;31:378–92.

Quadrant. C100F23-F1 product data sheet, Quadrant Plastic Composites; 2007. Available at: www.quadrantcomposites.com/files/cms1/DS_D100F23-F1.pdf. (Accessed July 2010).

Quadrant. Applications for building and construction, Quadrant Plastic Composites; 2010. Available at: atwww.quadrantcomposites.com/English/page_14aspx. (Accessed July 2010).

RCFL. Recycled Carbon Fibre Ltd, O_cial website at: www.recycledcarbonfibre.com; 2009.

Red C. Aerospace will continue to lead advanced composites market in 2006. Composites Manuf 2006;7:24–33.

Roberts T. The carbon fibre industry: global strategic market evaluation 2006–2010. Watford, UK: Materials Technology Publications; 2006. p. 10, 93–177, 237.

Rocky Mountain Institute. Carbon fiber cost breakdown, http://www.rmi.org/RFGraph-carbon_fiber_cost_breakdown; 2015.

Warnecke M, Wilms C, Seide G, Gries T, Yilmaz H, Lorz O. Chemical Fibres International, 2010; 60, 4, 213–14.

Warnecke M, Wilms C, Seide G, Gries T, Yilmaz H, Lorz O. Man-made fibre year book. Germany: Chemical Fibre International; November 201132 Supplement, 2011.

Warren. Development of low cost, high strength commercial textile precursor (PAN-MA), Oak Ridge, TN: Oak Ridge National Laboratory; June 2014. http://energy.gov/sites/prod/files/st099_warren_2014_o.pdf.

Warrior NA, Turner TA, Pickering SJ. AFRECAR and HIRECAR project results, In: Carbon fibre recycling and reuse 2009 conferenceHamburg, Germany: IntertechPir; 2009.

Watt W, Kelly A, Rabotnov YN. Handbook of composites. vol. I. Holland: Elsevier Science; 1985, pp. 327–87.

Wong KH, Pickering SJ, Turner TA, Warrior NA. Preliminary feasibility study of reinforcing potential of recycled carbon fibre for flame-retardant grade epoxy composite. In: Composites innovation 2007. Improved sustainability and environmental performanceBarcelona, Spain: NetComposites; 2007.

Xiaosong H. Fabrication and properties of carbon fibers. Materials 2009;2:2369–403.

Relevant websites

https://cs2.toray.co.jp.
https://omnexus.specialchem.com/news/industry-news/toray-increases-large-tow-carbon-fiber-production-capacity-000214244.
https://qbnnews.com/2019/07/22/carbon-fiber-market-kemrock-industries-hexcel-corporation-formosa-plastic-corporation/.
https://www.materialstoday.com/composite-processing/features/new-lease-of-life-for-cfrps/.
https://www.plasticstoday.com/recycling/recycled-carbon-fiber-key-mass-market-applications/7344798259925.
https://www.smithersapex.com/news/2016/november/carbon-fibre-market-forecast-to-grow-to-132,000-to.
http://www.formula1-dictionary.net.
http://www.futuremarketsinc.com.
http://www.nrel.gov.
http://www.smithersapex.com.

Further reading

Holmes M. Global carbon fibre market remains on upward trend. Reinforced Plastics 2015;58:38–45.
Meng F, McKechnie J, Turner T, Pickering SJ. Energy and environmental assessment and reuse of fluidised bed recycled carbon fibres. Compos Part A 2017a;100:206–14. https://doi.org/10.1016/j.compositesa.2017.05.008.
Meng F, McKechnie J, Turner T, Wong KH, Pickering SJ. Environmental aspects of use of recycled carbon fiber composites in automotive applications. Environ Sci Technol 2017b;51(21):12727–36. https://doi.org/10.1021/acs.est.7b04069.

CHAPTER 10

Future directions of the carbon fiber industry

In general the future for the carbon fiber industry looks extremely bright (Fitzer, 1990; Chung, 1994; Watt, 1985; Donnet and Bansal, 1990; Minus and Kumar, 2005, 2007; Fitzer et al., 1989; Hajduk, 2005; Xiaosong, 2009; http://www.zoltek.com/carbonfibre/the-future-of-carbon-fibre/). "Carbon fiber is already an important material in products across all markets and industries, and new applications and uses discovered every year. With its growth, carbon fiber and its composites offer great potential to the future of all markets and industries. With excellent strength to weight ratios and a long list of other benefits, carbon fiber's potential uses are currently growing in manufacturing, product design, energy conservation, natural resource conservation, the automotive industry, and more" (www.iarjset.com) (Table 10.1).

Some main areas of natural resource, automotive, and commercial industry-related growth comprise fuel-efficient vehicles, automotive parts and repairs, alternative energy, oil exploration, and construction and infrastructure.

The future of carbon fiber is commercial carbon fiber. The carbon fiber market demand increased gradually after the worldwide financial slump in 2009. The market is predicted to grow considerably in the next few years. This growth will take place in aerospace and defense, sporting goods, wind energy, automobiles, and even in civil engineering and electronics goods market. Global carbon fiber market size will grow by USD 1.74 billion (Consumer discretionary, Published reports 2019, https://www.technavio.com).

The carbon fiber-reinforced composite market is envisaged to increase to USD 36 billion in 2020 (https://theconversation.com/black-to-the-future-carbon-fibre-research-seeds-new-innovation-27148).

Aerospace and defense are the main drivers for growth in the carbon fiber market accounting for about a third of the total product market with regard to demand and about half of the total market revenue. This is predicted to grow till 2024 due to increasing air travel in Middle East, China, and other developing countries in the aerospace and defense industry. There is increasing use of carbon fiber in sports and leisure equipment. This is the second largest market with regard to revenue. Because of the rising consciousness about healthy standard of living and exercising, the requirement for sports utilities will have a vigorous expansion in the years to come. The increasing demand for light-bodied fighting aircrafts by the military will drive the industry in years ahead. Another major factor driving the carbon fiber market is wind energy segment approaching 15% of

Table 10.1 Future directions of carbon fiber industry.

Alternate energy
Wind turbines, compressed natural gas storage and transportation, fuel cells
Fuel-efficient automobiles
Currently used in small production and high-performance automobiles but moving toward large production series cars
Construction and infrastructure
Lightweight precast concrete, earthquake protection
Oil exploration
Deep sea drilling platforms, buoyancy, umbilical, choke, kill lines, drill pipes

Based on Fitzer E, Edie DD, Johnson DJ. In: Figueiredo JL, Bernardo CA, Baker, RTK, Huttinger KJ, editors. Carbon fibres filaments and composites. 1st ed. New York: Springer; 1989. p. 3–41, 43–72, 119–146; Hajduk F. Carbon fibres overview, In: Global outlook for carbon fibres 2005 Intertech Conferences, San Diego, CA; 11–13 October 2005; http://www.zoltek.com/carbonfibre/the-future-of-carbon-fibre/.

the worldwide demand. With exhaustion of fossil fuels, the world is now focusing on the production of energy from sustainable energy sources. The developments in the wind energy segment will result in growing use of carbon fiber in the production of lightweight and strong turbine blades and their parts. Another main segment that will push the demand for carbon fiber is the automotive industry. With increasing safety apprehensions, consumers are now focusing on strong automobile bodies from traditional metal bodies. Improving the life quality in the developed countries will impel the demand for stylish sports and also the luxury cars. This will drive the carbon fiber market to a large extent. The main types of carbon fibers based on the resin type are thermoset and thermoplastic fibers. Most of the aerospace prepregs are produced from thermoset fibers because of its mechanical advantage. The shelf life of thermoplastics is higher. These are finding more use in areas where the need of strength and resistance is surpassed by toughness (www.gminsights.com).

> Carbon fiber finds major use in aerospace and defense industry, wind energy sector, sports equipment industry, automotive industry and others. Aerospace industry is the main end user, of which commercial aerospace sector has the major market share in terms of volume and also revenue. The wind energy segment has a lot of promise for the market growth because of increase in global renewable energy demand. Light weight and strong gym equipment and other luxury items will increase the global demand for carbon fiber by 2024. With the focus of leading automobile manufacturers shifting towards weight reduction of the cars, carbon fiber-reinforced plastics will find an increasing demand in the coming years. Among other industries there are the civil engineering industry and electronics goods. With growing population and upgrading social status of the people around the world, the demand for light weight and cost effective electronic goods will affect the carbon fiber market positively.
>
> *(www.gminsights.com)*

North America is the most demanding region for the industry having yearly production capacity more than third of the carbon fiber market globally. With more R&D in the aerospace and defense segments in the United States, this demand would grow in the near future. Europe is the second leading producer with its commercial and also military aircraft producing units. In Germany and Italy the large number of automotive industries will also affect the carbon fiber market (www.gminsights.com).

Japan ranks fifth in terms of production capacity in Asia. Because of its disaster-prone landmass, the demand for lightweight construction equipment will increase the use of carbon fiber in the coming years. The dominant producer of carbon fiber was Toray Industries and produced over 20,000 tons of carbon fiber. Other competitive producers are Zoltek, Toho Tenax, Mitsubishi Rayon Co. Ltd., SGL Carbon SE, Formosa Plastics Corp., Hexcel Corporation, and others.

Although there has been a continuous growth of carbon fibre production over the last few years, large-volume applications of carbon fibres, have been impeded by the high fibre costs and the lack of techniques for the high-speed production of composites. However, present investigations and future developments might well change the market and establish carbon fibres as a mass product similar to other synthetic fibres or even metals. Renewable raw materials are particularly interesting sources of carbon fibres.

(Bajpai, 2017; Baker and Rials, 2013; Bajpai, 2013; Frank et al., 2014)

To fully develop Carbon Fibre in different industries, Carbon Fibre manufacturers need to continue to increase their capacity and change their mindset so that they are committed to commercialisation. The perfect conditions that would allow the Carbon Fibre industry to reach its vast potential are if carbon manufacturers:

- *Target new applications.*
- *Develop new and lower-cost technology.*
- *Reinvest profits with long-term objectives in mind (no small operators focusing on low volume and high price).*
- *Fully understand the costs and future strategy of suppliers.*
- *Identify and focus on market drivers.*
- *Work to aggressively reduce costs.*
- *Consolidate so that weaker players help strengthen the stronger ones.*
- *Share incremental improvements to help support market growth.*
- *Understand that the primary competitors to Carbon Fibre are other materials (not other Carbon Fibre manufacturers).*

(http://www.zoltek.com)

Carbon fiber will be used in flywheels. These are developed for the storage of wind energy during nonpeak generation. One flywheel consumes essentially 1100 lb./500 kg of carbon fiber. It is used in the natural gas vehicle market for gas tanks. There are 11–12.5 million natural gas vehicles on the road, and this is expected to increase to 65 million natural gas vehicle by 2020, with demand for 166.5 million pressure vessels. In the automotive sector, carbon fiber has developed a niche in the "supercar" segment, where its high cost is not a hindrance to sale (http://nextbigfuture.com). "Civil engineering (example bridges and other infrastructure) as well as the oil and gas industries have been resistant to carbon (at least in part) because they too tend to be change-averse and find greater comfort in their highly regulated environments with legacy steel and concrete technologies" (www.compositesworld.com).

Ford company has become the latest major producer and has collaborated with a technology company to investigate the use of carbon fiber products in its vehicles (http://www.caradvice.com.au/167855/ford-exploringcarbonfibre-for-future-vehicles/).

The agreement with Dow Automotive Systems is the second part of two-pronged endeavor of Ford for reducing the energy use of its fleets, thus complementing its downsizing of the EcoBoost engine and efficiency strategy (www.carlookout.com.au).

Ford is expecting the use of carbon fiber components to have a significant role in its goal for reducing the weight of its new vehicles by ≤ 750 lbs. (340 kg) by 2020. It is examining many new materials, design processes, and manufacturing methods, which will satisfy rising quality and safety standards and also reducing weight (www.carlookout.com.au).

Partnership will look to combine the design, engineering, and high-volume vehicle production experience of Ford with the strength in R&D, materials science, and high-volume polymer processing of Dow Automotive. If the corporation succeeds, carbon fiber components could be used in its vehicles before the end of the decade as it intends to improve the efficiency, reduce the emission, and extend range for its plug-in hybrid and electric vehicles.

The move is following commitments from many other international producers for investing in carbon fiber technology as a strategy for saving weight. BMW has strengthened its partnership with the carbon fiber producer SGL Group. This group is at the forefront of launching the electric and plug-in hybrid i3 and i8 models, which will use a higher percentage of the lightweight and higher strength material in their construction.

Mercedes-Benz has partnered with Toray—the Japanese carbon fiber producer—whereas Audi has also tied up with German manufacturer Voith (Car News, Ford https://www.caradvice.com.au).

Significant hurdles remain before high-volume usage of carbon composites become a reality. However, the outlook is, in general, positive, with significant progress being made through various industry-wide and academic initiatives. With commercial and technical issues resolved, carbon fiber is expected to be a key enabler in the drive for greater fuel efficiency.

Researchers are trying to ascertain how to reduce the cost of carbon fiber composites. Oak Ridge National Laboratory (ORNL) along with Tennessee University is optimizing the raw materials and is trying to find out alternative forms of carbon fiber from renewable resources. A favorable item is lignin, a waste produced during the pulping process.

Developing an effective carbon fiber oxidation also significantly lowers the cost of this raw material. Plasma processing technology aids the rapid oxidation of precursor fibers.

ORNL is collaborating with Atmospheric Glow Technologies. This company has expertise in atmospheric pressure plasma processing for producing and using plasmas (http://www.ornl.gov/info/press_releases/get_press_release.cfm?ReleaseNumber=mr 20060306-00). The project shows that an appropriately designed microwave-assisted plasma energy delivery system can quadruple production rate and reduce energy requirements and carbon fiber price by ≤20%. For evaluating these innovative processes on a comparable basis, ORNL has established a modular carbon fiber research line. ORNL has installed an advanced performing system that possesses a robotically activated arm that chops and sprays fiber and a binder in powder form for developing fiber preforms. These preforms are injected along with resin in a mold at high temperature and are fused under pressure for creating the final part. As a result the fibers glue together and provide remarkable strength. The preforming process is the initial step in producing the carbon fiber, which is lightweight, durable, and safe and could solve the oil crisis.

> *Lignin-based carbon fibre is the most value-added product from a wood-based biorefinery and has the potential of providing a low-cost alternative to petroleum based precursors to manufacture carbon fibre, which can be combined with a binding matrix for producing a structural material with much greater specific strength and stiffness compared to conventional materials such as aluminium and steel. The market for carbon fibre is projected to grow exponentially to fill the requirements of clean energy technologies such as wind turbines and to improve the fuel economies in vehicles through lightweighting. In addition to cellulosic biofuel production, lignin-based carbon fibre production coupled with biorefineries may provide $2400–$3600 added value dry Mg^{-1} of biomass for vehicle applications (Langholtz et al., 2014). Compared to production of ethanol alone, the addition of lignin-derived carbon fibre could increase biorefinery gross revenue by 30–300%. Using lignin-derived carbon fibre in 15 million vehicles per year in the United States could reduce fossil fuel consumption by 2–5 billion litres $year^{-1}$, reduce carbon dioxide emissions by about 6.7 million Mg $year^{-1}$, and realize fuel savings through vehicle lightweighting of $700–$1600 per Mg biomass processed. The value of fuel savings from vehicle lightweighting becomes economical at carbon fibre price of $6.60 kg^{-1} under current fuel prices, or $13.20 kg^{-1} under fuel prices of about $1.16 l^{-1}.*
>
> *(Bajpai, 2017)*

Several problems are faced in providing lignins appropriate for manufacturing carbon fiber. Till now, suitable lignins have not been obtained, which can be processed into carbon fiber having desired strength characteristics. This is due to the nonavailability of lignins with right properties so that they can be melt spun into fiber and processed to carbon fiber at reduced cost. This needs a low T_g lignin to be melt spun, a high T_g lignin to assure low cost, and reduced cost of processing of the lignin before fiber spinning.

Using other fiber spinning methods that are not bound by the severe technical needs required for the conventional melt spinning of fibers, cost reduction can be achieved. Nonetheless the provision of refined or improved lignins may yet provide the required lignin performance qualities needed for traditional melt spinning or certainly the use of any fiber forming method for producing carbon fiber of superior strength properties (Baker et al., 2012; Morck et al., 1986). Considering the processes used for manufacturing lignin, lignin products will need purification and refining to be appropriate for use in the production of carbon fiber. This is found to be correct for even a highly developed organosolv processes, which produces lignins of higher purity and with much more appropriate T_g and T_s properties in comparison with the competing organosolv techniques (Black et al., 1998; Bozell et al., 2011). In the connection of lignocellulosic biorefining for producing biofuels, the use of organosolv techniques are cost effective with pretreatment and other fractionation techniques and with the added provision of comparatively pure lignins (Bozell, 2010a,b, 2011; Michels and Wagemann, 2010; Black et al., 1998; Lignol Energy Corporation, 2010, 2011). However, for performing a particular type of pretreatment, fractionation, or lignin refining methods for achieving specific lignins for production of carbon fiber, better understanding of lignin and its conversion to materials are needed (Baker and Rials, 2013).

India is a potential place for growth of carbon fiber industry. Nevertheless, its use has been restricted due to the unavailability of carbon fibers commercially nearby. Defense and aerospace segments have been largest users of carbon fibers so far. But they are now moving from R&D to production phase.

In the civilian sectors
(a) *India is poised to produce ~50,000 MW of Wind Energy by 2018. Hence will require additional 4000–5000 MTS of carbon fibres,*
(b) *India is one of the seven main countries of the world with Natural Gas Vehicles (NGV). As of now, number of such vehicles is only ~1.6 million (This number is only 4% of the total vehicles) and is poised to grow. Conversion of only 10% of CNG steel cylinders to CFRP cylinders per annum will generate additional demand of ~1000 MT/year.*

It is pertinent for the International Carbon Fibre Community to take note of the above mentioned facts. It is a ripe place to set up a manufacturing unit of a suitable capacity, keeping in mind that world class scientific manpower (in the area of carbon fibres) is available locally.

As always, cost of CFRP is the challenge. This challenge is both from within and also from competing materials—typically metals and alloys. Reducing cost of carbon fibres is a challenge from within. In case of polyacrylonitrile based carbon fibres, costing of carbon roughly runs as;

Precursor ~50%
Stabilization and carbonization ~37% plus others.

(www.gocarbonfibre.com)

Not much can be done in case of polyacrylonitrile precursor cost. Completing carbonization process has been achieved in 5 min only—hence resulting in saving of gas, energy, and time. There are 1000 million vehicles on the planet today. This number would grow to 3000 million by 2030 and 3000 million by 2050. In the BRIC countries the number of vehicles is increasing. It should be noted that the use of petroleum by vehicles will increase from 51% to more than 60% of the total petroleum production. This is unsustainable. People can afford to have more vehicles by the use of CFRP in automobiles—making them lighter and thereby reducing the fuel consumption. India is the fourth largest economy of the world having a very strong automobile manufacturing hub (www.gocarbonfibre.com).

Wind power is another large user of carbon fibers. Both these areas, which are expected to drive the use of carbon fibers in the next decade, will use heavy tow. So, future development of carbon fiber production will take place in heavy tow.

About 30% of total energy is used by transport alone, that is, by cars, trucks, and buses. Out of the total transport consumption, ~75% is used by cars and trucks alone.

So far, liquid fuel cannot be replaced in transport sector since gasoline has 60–300 times more energy in comparison with the electric batteries per unit mass and can be easily stored, transported, handled, etc.

Nevertheless, the harmful consequences of liquid fuel are so overweighing that the transport with liquid fuel is unsustainable. Furthermore, due to superior performance, the share of lithium-ion battery is ~40% out of the total market of all batteries.

All new products/materials are basically developed for their most sophisticated applications to start with (carbon fiber is a good example). Nonetheless the products/materials will become a success on a commercial scale if applications on a large scale are realized.

Lithium-ion battery market is expected to grow exponentially mostly due to its application in automobile sector. Within this sector the light vehicles and light commercial vehicles will be used. China and Europe would lead the market by 2020, thus grabbing the lead from Japan.

Lithium-ion battery business is identified as a strategic industry, so much that the United States has sanctioned > $2.5 billion for this area to remain important considering the threats from Europe and China.

Some barriers for growth of carbon usage in transportation include the following:
- Cost of the fibers.
- Fiber availability.
- Design methods—predictive modeling of carbon fibers is quite complex and difficult, particularly while designing for crash critical applications. Test standards are not consistent.
- Processing technologies—many composite processing techniques are optimized for performance, not production rate efficiency.
- Larger structures are more cost sensitive to the cost of raw material. Less material is required for smaller structures.

- Meeting requirement of mass production of vehicles.
- Surface treatment and sizing technologies are guarded by industry.
- The lack of resin targeted sizing systems.
- Property translation is immature for carbon fiber in automotive resins.
- Product form.

The future of carbon fiber is very bright, with immense potential in several industrial applications. Some key sectors include the following:

- Alternate energy—CNG storage, wind turbines, transportation, and fuel cells.
- Fuel-efficient automobiles—presently used in small production and high-performance automobiles, but moving toward large-production series cars.
- Construction and infrastructure—lightweight precast concrete and earthquake protection.
- Oil exploration—deep sea drilling platforms, buoyancy, umbilical, choke, kill lines, and drill pipes (www.carbonfiber-products.com).

The key question to be considered is what would be the price of carbon fiber to be a high-volume material option? Primary competition to carbon fiber is not from steel as it was in the past, but from aluminum and magnesium. The cost of producing carbon fibers has been reduced considerably in the last two decades, and the scientists are bringing the cost down every day. With development of low-cost precursors and efficient processing technologies, the cost target for carbon fiber can meet the requirement of industry. Carbon fiber applications in the automotive industry are predicted to grow 10%–15% a year. As they do, several of the applications that were impossible will become a reality. Carbon fibers are used scarcely in automotive applications, but sooner or later, whole body panels may be made from them. Commercial production of carbon fiber could become one of the very large growth segments in the automotive industry in the second half of this decade. India is bestowed with good natural resources, and research efforts can be diverted for developing low lignin, rayon, and polyolefin-based carbon fibers.

India's first carbon fiber manufacturing unit for catering to aerospace and defense sectors is being set up by Reliance Industries (https://www.livemint.com/Companies/zEbKXK6O3FgD0Jdtg0y5jM/Reliance-investing-in-Indias-first-carbon-fibre-unit.html). Reliance is the owner of largest oil refining complex in the world. It will also produce high-volume and low-cost composite products such as modular toilets, homes, and composites for windmill blades and rotor blades. Reliance is developing 3D printing of broad range of plastic and metal products. Reliance is developing new business verticals in the petrochemical business for capturing Rs 30,000 crore composites market and is planning to produce graphene, superior plastics and elastomers, and fiber-reinforced composites, which can replace steel (http://bizshorts.in).

It acquired Kemrock Industries and is focusing on thermoset composites like glass and carbon fiber-reinforced polymers (FRPs). The ability to impart extraordinary strength (similar to or better than steel) at an appreciably reduced weight is an important feature of FRPs. Furthermore, FRPs can tolerate severe weather, have a long life with reduced maintenance, are corrosion resistant, and can be molded into any shape. Composites are used in a broad range of applications: industrial, railways, renewable energy, defense, and aerospace. RIL is expecting the newly started Reliance Composites Solutions business to be the No. 1 in India in the field of composites. It has plans to focus on design and specification-driven applications that have the potential for yielding better returns. The focus areas include wind mill blades and parts for railways and metros, which have demanding standards of performance and safety (particularly fire retardant). Also on the radar are carbon wraps to restore old infrastructure of India for enhanced seismic performance and pipes. "RCS will design and administer low-cost and high-volume products such as modular toilets and homes to support the Swachh Bharat Mission, disaster relief measures and Housing for All programmes initiated by the Indian Government." According to RIL, industrial 3D printing (particularly with metal) is reaching an inflection point, and the company is developing the competence for designing and printing a broad range of products using 3D printing technology in plastic and metal from prototypes to functional parts (www.financialexpress.com).

For composites, carbon fiber presents several performance and lightweighting advantages. Reduced production cost and higher use of recycled material are widening the applications that might benefit from its properties. Higher cost has been a main obstruction to extensive use of carbon fiber as a strong, rigid reinforcement for advanced composites. One scheme in progress is reducing these costs. In the United States, scientists at Oak Ridge National Laboratory (ORNL) have developed a method that can reduce the cost of carbon fiber by up to 50% and energy consumption by more than 60%. ORNL's new method, developed at its Carbon Fiber Technology Facility, took more 10 years of research in the area. This would speed up the use of carbon fiber composites in high-volume industrial applications. After rigorous analysis and successful prototyping by industrial partners, ORNL developed a new method for licensing. According to Dr. Alan Liby, deputy director of the Advanced Manufacturing Program at ORNL, "Through a competitive selection process, ORNL is working to negotiate up to five license agreements for its low-cost carbon fiber process" (www.elgcf.com).

> *LeMond Composites was the first to sign a license agreement and other licenses are still in negotiation. The licensees range from start-ups to established players in the carbon fiber production field. Partners will be selected based on their capabilities, business plans,*

and commitments to manufacture in the United States. Expectations are to see this technology in the marketplace by 2018 and the licensees will explore additional market opportunities.

(cfdriveshaft.com)

The innovation is about the production of low-cost carbon fiber. The properties of the material that the licensees produce will establish the final use. The carbon fiber manufactured by ORNL meets the performance criteria prescribed by some automotive producers for high-strength composite materials used in high-volume applications. Nevertheless, the process would speed up the use of advanced composites in other manufacturing applications. Future markets could include applications in the wind turbine and gas storage industries. The technology was developed at the Carbon Fiber Technology Facility with a capacity of up to 25 tons per year. The licensees will increase that capacity in their own operations (www.materialstoday.com).

The carbon fiber industry will reach its vast potential if carbon fiber manufacturers:
- *Target new applications*
- *Develop new and lower cost technology*
- *Reinvest profits with long term objectives in mind, essentially eliminating the low volume, high price mentality.*
- *Fully understand supplier's costs and future strategy*
- *Identify and focus on market driver's*
- *Work to aggressively reduce costs*
- *Consolidate so that weaker players help strengthen the stronger ones*
- *Share incremental improvements to help support market growth*
- *Understand that the primary competitors to carbon fibers are other materials, not other carbon fiber manufacturers.*

(http://www.zoltek.com)

As there is a growing demand for carbon fiber around the globe, several manufacturers are making investment to increase the capacity and improvements. SGL and BMW have increased their annual production and expanded their plants, dealing a joint venture; Mitsubishi Rayon is also planning to increase the capacity by two times; Ford and DowAksa are jointly doing R&D; and Hexcel has invested in a new factory for carbon fiber production, besides building up a new precursor (Witten et al., 2016, 2017; iopscience.iop.org).

References

Baker DA, Gallego NC, Baker FS (2012). On the characterization and spinning of an organicpurified lignin toward manufacture of low-cost carbon fibre. J Appl Polym Sci 227–234.
Baker DA, Rials TG. Recent advances in low-cost carbon fibre manufacture from lignin. J Appl Polym Sci 2013;130:713–28.
Bajpai P (2013). Update on Carbon Fiber Smithers Rapra, UK
Bajpai P. Carbon fibre from lignin. In: Springer Briefs in material science. Switzerland: Springer (Springer Nature); 2017.

Black SK, Hames, BR, Myers MD (1998). U.S. Patent 5,730,837, assigned to Midwest Research Institute.
Bozell JJ. Connecting biomass and petroleum processing with a chemical bridge. Science 2010a;329:522–3.
Bozell JJ. Fractionation in the biorefinery. BioResources 2010b;5(3):1326–7.
Bozell JJ, O'Lenick C, Warwick S (2011) Biomass fractionation for the biorefinery: heteronuclear multiple quantum coherence-nuclear magnetic resonance investigation of lignin isolated from solvent fractionation of switchgrass. J Agric Food Chem 59:9232–9242.
Chung DL. Carbon fibre composites. Boston: Butterworth-Heinemann; 1994. p. 3–11.
Donnet JB, Bansal RC. Carbon fibres. 2nd ed. New York: Marcel Dekker; 1990. p. 1–145.
Fitzer E. Figueiredo JL, Bernardo CA, RTK B, Huttinger KJ, editors. Carbon fibres filaments and composites. Dordrecht: Kluwer Academic; 1990 p. 3–4, 43–72, 119–146.
Fitzer E, Edie DD, Johnson DJ. Figueiredo JL, Bernardo CA, Baker RTK, Huttinger KJ, editors. Carbon fibres filaments and composites. 1st ed. New York: Springer; 1989 p. 3–41, 43–72, 119–146.
Frank E, Steudle LM, Ingildeev D, Spörl JM, Buchmeiser MR (2014) Carbon fibres: precursor systems, processing, structure, and properties. Angew Chem Int Ed 53:2–39.
Hajduk F. Carbon fibres overview, In: Global outlook for carbon fibres 2005 Intertech Conferences, San Diego, CA; 11–13 October 2005.
Lignol Energy Corporation. Lignol announces biorefining technology breakthrough with AlcellPlusTM, https://www.lignol.ca/news/AlcellPlus_Press_Release_FINAL_Oct_19_2010.pdf; 2010 Company Press Release, 19 Oct 2010.
Lignol Energy Corporation. Lignol develops first renewable chemical product. Company press release, 11Apr 2011;2011.
Michels K. Wagemann. The German lignocellulose feedstock biorefinery project. Biofuels Bioproducts & Biorefining 2010;4:263–7. 2010.
Minus ML, Kumar S. The processing, properties, and structure of carbon fibers. JOM 2005;57(2):52–8.
Minus ML, Kumar S. Carbon fibre. Kirk-Othmer Encycl Chem Technol 2007;26:729–7495.
Morck R, Yoshida H, Kringstad KP, Hatakeyama H (1986) Fractionation of kraft lignin by successive extraction with organic solvents. Holzforschung 42:111–116.
Watt W. Kelly A, Rabotnov YN, editors. Handbook of composites. vol. I. Holland: Elsevier Science; 1985. p. 327–87.
Witten E, Kraus T, Kühnel M. Composite market report. Germany: Industrievereinigung Verstärkte Kunstsoffe; 2016.
Witten E, Sauer M, Kühnel M. Composites market report. Germany: Industrievereinigung Verstärkte; 2017.
Xiaosong H. Fabrication and properties of carbon fibers. Materials 2009;2:2369–403.

Relevant websites

bizshorts.in.
Carbon fibre cars could put us on highway to efficiency, http://www.ornl.gov/info/press_releases/get_press_release.cfm?ReleaseNumber=mr20060306-00.
Car News, Ford. https://www.caradvice.com.au.
cfdriveshaft.com.
Consumer discretionary. Published reports, https://www.technavio.com; 2019.
Ford exploring carbon fibre for future vehicles, http://www.caradvice.com.au/167855/ford-exploring carbonfibre- for-future-vehicles/.
Future of carbon fibre, http://nextbigfuture.com/2010/08/future-of-carbon-fibre.html.
http://nextbigfuture.com.
http://www.caradvice.com.au/167855/ford-exploringcarbonfibre-for-future-vehicles/.
http://www.ornl.gov/info/press_releases/get_press_release.cfm?ReleaseNumber=mr20060306-00.
http://www.zoltek.com/carbonfibre/the-future-of-carbon-fibre/.
https://theconversation.com/black-to-the-future-carbon-fibre-research-seeds-new-innovation-27148.
https://www.livemint.com/Companies/zEbKXK6O3FgD0Jdtg0y5jM/Reliance-investing-in-Indias-first-carbon-fibre-unit.html.

http://www.carbonfiber-products.com.
http://www.carlookout.com.au.
http://www.compositesworld.com.
http://www.elgcf.com.
http://www.financialexpress.com.
http://www.gocarbonfibre.com.
http://www.gminsights.com.
http://www.iarjset.com.
http://www.materialstoday.com.
http://www.zoltek.com.

CHAPTER 11

Future research on carbon fibers

Carbon fiber production is rapidly increasing to cope with market demand. As the production costs are steadily falling, raw material producers are expanding their production capacity. And this will make it possible to launch lower-cost carbon fibers to the market while maintaining the quality levels.

> *Carbon fiber is surprisingly versatile, not only in the material itself but also in the way it can be used. It is possible to create a carbon fiber-based construction where the stresses vary in different places to support what it is being used for without incurring weight penalties.*

(https://www.themanufacturer.com/articles/five-reasons-why-carbon-fibre-is-the-future-of-design/)

Future research on carbon fiber will focus on cost reduction and improving the property. The two most important carbon fiber precursors are polyacrylonitrile and mesophase pitch (Chung, 1994; Donnet and Bansal, 1990; Minus and Kumar, 2005; Minus and Kumar, 2007; Fitzer et al., 1989; Hajduk, 2005).

A considerable amount of work has been conducted on relating fiber structure to properties and translating the relationship into production to reduce the cost of production or increase the properties of fiber (Xiaosong, 2009). However, several challenges such as reduction in cost, improving the tensile and compressive strength, and the development of alternative precursors remain. Optimizing the microstructure of carbon fiber may enhance carbon fiber strength by reducing its flaw sensitivity. The microstructure of carbon fiber is dependent upon morphology of the precursor and processing conditions. R&D in these areas could help in the development of carbon fiber showing better performance (Xiaosong, 2009; www.mdpi.com).

The mechanical properties of carbon fiber rely heavily upon its microstructure (Minus and Kumar, 2007). The tensile, flexural, and shear strength of pitch carbon fiber can be improved by randomizing the graphite distribution in the transverse direction of the fiber. The size and distribution of crystallites can be changed by changing the design of die, spinning temperature, spinning rate, and the polyacrylonitrile copolymer type and its stereoregularity. But research in this area is not ample. Optimization of the size, shape, and distribution of crystallites by changing processing parameters and their effect on fiber properties should be explored. Understanding of the relationship between the ratio of sp^2/sp^3 bonding and fiber properties could also lead to better fiber properties (www.mdpi.com).

Preliminary research on reducing the processing cost by melt spinning of polyacrylonitrile has shown encouraging results. Carbon fibers with a tensile strength of ≈3.6 GPa and a modulus of ≈233 GPa have been produced by melt-assisted spinning process (www.tandfonline.com). More improvement on the mechanical properties of fibers has been hampered by defects. Use of different plasticizers (besides water) and modification of the process for reducing the defects produced by the evaporation of plasticizers need to be examined.

In comparison with polyacrylonitrile, pitch is a low-cost precursor. But the processing cost is high because the isotropic pitch needs to be converted into a mesophase pitch for producing high-performance carbon fiber. Researchers are trying to reduce the cost of preparation of mesophase pitch. A breakthrough in this area could substantially reduce the cost of pitch carbon fiber.

Several polymers have been examined as low-cost precursor. Lignin is a potential precursor for low-cost and medium properties carbon fiber. Organosolv lignin of high purity can be directly spun into precursor fibers. Another low-cost precursor showing promise is polyethylene. The carbon fiber made from polyethylene has shown a high carbon yield of 75% and a strength of 2.5 GPa. But not much work has been done on process optimization. Research to reduce the stabilization time using different methods, for instance, addition of active chemicals in the spinning dope, should be examined. The effect of heat stretching on the mechanical properties of the resultant carbon fiber also should be studied. The use of carbon fiber-based composite materials will become commonplace when materials such as steel and aluminum need to be replaced (Fitzer et al., 1989; Hajduk, 2005). However, to do so, manufacturing on a large scale will be needed, and nanoparticles will find their way into matrix materials, where they will improve toughness. Carbon nanotube–based precursors will be used for producing the next generation of carbon fiber materials. These materials have elastic moduli >500 GPa (considerably higher than the 300 GPa of existing materials). A major factor of this processing technology is that it will use equipment, so plants presently geared up to produce carbon fiber from polar will be able to use this new chemistry without alteration. These next-generation carbon fiber will no doubt find application in aircraft, where their high strength and lightweight properties will result in more fuel-efficient aircraft.

Significant hurdles remain before high-volume usage of carbon composites becomes a reality. However, the outlook is, in general, positive, with significant progress being made through various industry-wide and academic initiatives. With resolution of commercial and technical issues, carbon fiber is expected to be a key enabler in the drive for greater fuel efficiency. The most recent developments in carbon fiber technology are small carbon tubes called nanotubes. These hollow tubes, some as small as 0.00004 in (0.001 mm) in diameter, have exceptional mechanical and electrical properties that may be helpful in producing new high-strength fibers, submicroscopic test tubes, or perhaps new semiconductor materials for integrated circuits (www.materialsciencejournal.org).

Following three predictions have been made about the future of carbon fiber industry that will play an important role in the explosive growth (www.rockwestcomposites.com):

1. Reduction in cost

At this time, there is no debate that the cost of carbon fiber composite materials for fabrication is high. Most of the carbon fiber composites are produced by machines, but the process appears to be very laborious.

At present, there is a trend to reduce the processing cost of composites by increasing mechanization. The development of better and advanced machines for producing everything from tubing to plates to prepreg materials is being seen. As Henry Ford made the automobile reasonably priced by using assembly line production, our industry is going to make carbon fiber composites more profitable through mechanization (https://www.rockwestcomposites.com/blog/3-predictions-about-the-future-of-carbon-fiber-composites/.).

2. Increase in versatility

Theoretically, carbon fiber composites can be used to produce just about anything. Some things are easier to manufacture in comparison with others. A carbon fiber tube can be fabricated very easily than an airplane wing, for instance. But that is changing.

With new investigation being conducted on nanocomposites and closed mold infusion, our industry is finding out pioneering ways for using composites that can meet several requirement with regard to strength, durability, size, and scope. The versatility of carbon fiber is expanding rapidly as our understanding of composite materials is growing. This would lead the way into a future of electric cars, bigger airplanes, and more.

3. Improvement in aesthetics

At the moment, one can buy carbon fiber materials from Rock West Composites in black. People often ask for more colors. For this, investment in alternative weaving styles and hybrid wovens that include nylons, Kevlars, fiberglass, and even some naturally grown flax fibers is being made. For selling the composites the appearance has to be good, and efforts are being made to that end.

Recently, Chinese researchers are using structural colors instead of pigmented dyes for coloring carbon fiber fabrics. Their discovery is expected to have a significant effect on aesthetics once it hits the market.

> *At present, the automotive industry is accounting for less than 1% of world carbon fiber output, but the trend is to increase this number. Carbon fiber–reinforced composites are known to be up to 35% lighter than aluminum and 60% lighter than steel. In a time when government policies have required a reduction of exhaust emissions from cars, replacement of automotive components for weight reduction is presented as an interesting alternative. For a long time, carbon fibers were not considered as a hypothesis because of their cost; however, lignin-based carbon fiber has been proven to be a*

solution to overcome this limitation. Considering the mechanical properties obtained so far, the automotive industry is considered the best segment market to introduce lignin-based carbon fibers.

(http://iopscience.iop.org)

References

Chung DL. Carbon fibre composites. Boston, MA, USA: Butterworth-Heinemann; 1994. p.3–11.
Donnet JB, Bansal RC. Carbon fibres. 2nd ed. New York: Marcel Dekker; 1990. p.1–145.
Fitzer E, Edie DD, Johnson DJ. In: Figueiredo JL, Bernardo CA, Baker RTK, Huttinger KJ, editors. Carbon fibres filaments and composites. 1st ed. New York: Springer; 1989. pp. 3–41, 43–72, 119–146.
Hajduk F. Carbon fibres overview, In: Global outlook for carbon fibres 2005 Intertech Conferences San Diego, CA, 11–13 October 2005; 2005.
Minus ML, Kumar S. The processing, properties, and structure of carbon fibers. JOM 2005;57(2):52–8.
Minus ML, Kumar S. Carbon fibre. Kirk-Othmer Encycl Chem Technol 2007;26:729–7495.
Xiaosong H. Fabrication and properties of carbon fibers. Materials 2009;2:2369–403.

Relevant websites

https://www.rockwestcomposites.com/blog/3-predictions-about-the-future-of-carbon-fiber-composites/.
https://www.themanufacturer.com/articles/five-reasons-why-carbon-fibre-is-the-future-of-design/.
http://www.rockwestcomposites.com.
http://iopscience.iop.org.
http://www.mdpi.com.
http://www.materialsciencejournal.org.
http://www.tandfonline.com.

Further reading

Bajpai P. Carbon fibre from lignin, SpringerBriefs in Material Science. Springer (Springer Nature); 2017.

Appendix: Carbon nanotubes

Carbon nanotubes (CNTs) were discovered in 1991. CNTs belong to the fullerene family. They are made of carbon allotropes. The atoms are linked in the shape of cage-like structures, for instance, a hollow sphere and ellipsoid (Kroto and Fischer, 1993). Fullerenes consist of graphene sheets of linked hexagonal and pentagonal rings, which impart them their curved structure (Dresselhaus et al., 1996a,b). Graphene consists of a single layer of carbon atoms that are arranged in a two-dimensional hexagonal lattice. It is a semimetal having an overlapping between the valence and conduction bands, that is, it has a zero bandgap (Tanaka and Iijima, 2014). "The buckminsterfullerene (buckyball/C60) is the most widespread spherical fullerenes. It is a nanoscale molecule containing 60 carbon atoms. Each atom is bonded to three other nearby atoms and form hexagons and pentagons, with the ends curved into a sphere. The C70 molecule is another spherical fullerene which is chemically stable. In addition, other smaller metastable species, like C28, C36, and C50, have been found. Fullerenes have been present in nature for a long time; small amount of fullerenes in the form of C60, C70, C76, C82, and C84, have been found hidden in soot. Nanotubes consist of sp^2-hybridized carbon bonds. These bonds are tougher than the sp^3-hybridized carbon bonds found in diamond, thus making for the excellent strength and rigidity of nanotubes. In addition, they have extremely high electrical conductivity, high charge carrier mobility, high chemical stability, large specific surface ratio, high aspect ratio, outstanding mechanical properties, and heat conductivity, with some SWCNTs showing superconductivity, high charge carrier mobility" (nanoscalereslett.springeropen.com; Kroto and Fischer, 1993; Kroto et al., 1985; Fischer et al., 1997; Coleman et al. 2006; Martel et al., 1998; Frank et al., 2010; Abdalla et al., 2015; Peigney et al., 2001; Wang et al., 2009; Li et al., 2000, 2002a, 2010; Demczyk et al., 2002; Hone et al., 2002; Tang et al., 2001; Ferrier et al., 2004; Bethune et al., 1993). These properties are making CNTs an important subject in nanoscience and electronics research (nanoscalereslett.springeropen.com; Colomer et al., 2002; Wilson et al., 2002; Harris, 1999; Saito et al., 1998; www.azonano.com; Dresselhaus et al., 1996a,b, 2000; Ebbesen 1994, 1996; Yakobson and Smalley, 1997; Ajayan, 1999; Reich et al., 2004; Saito, 1998; Meyyappan, 2004; Iijima and Ichihashi, 1993; Ebbesen and Ajayan, 1992, Wei et al., 2001, Gao et al., 2000; Dupuis, 2005; Javey et al., 2003).

Nanotubes are cylinders that do not fold around to create a sphere. CNTs consist of carbon atoms that are linked in hexagonal shapes. Each carbon atom is joined to three other carbon atoms with covalent bond. The diameters of CNTs are as small as 1 nm, and the length is up to several centimeters. The buckyballs and CNTs are strong and are not breakable. They can be curved, and they will come back to their original shape when released (www.understandingnano.com). The shape of buckyballs is spherical, whereas a nanotube has cylindrical shape and has at least one end normally capped with a fraction of the buckyball structure (Lee et al., 2015; https://www.sciencedaily.com/terms/carbon_nanotube.htm).

The shape of carbon nanotube is cylindrical with open ends (Fig. A.1) or has closed ends, formed by a few of the carbon atoms combining into pentagons on the end of the nanotube (Fig. A.2) (He et al., 2013).

CNTs have novel properties that can be used in different types of applications in nanotechnology, electronics, optics, and other fields of materials science. They show

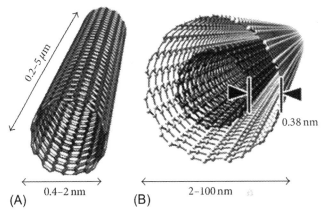

Fig. A.1 Conceptual diagrams of single-walled carbon nanotubes (SWCNT) (A) and multiwalled carbon nanotubes (MWCNT) (B). *Reproduced with permission from He H, Pham-Huy LA, Dramou P, Xiao D, Zuo P, Pham-Huy C. Carbon nanotubes: applications in pharmacy and medicine. Biomed Res Int 2013;2013:578290.*

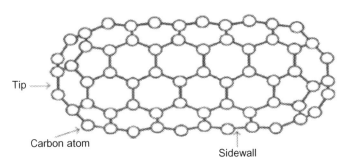

Fig. A.2 A carbon nanotube with closed ends. *Reproduced with permission from He H, Pham-Huy LA, Dramou P, Xiao D, Zuo P, Pham-Huy C. Carbon nanotubes: applications in pharmacy and medicine. Biomed Res Int 2013;2013:578290.*

high strength and distinctive electrical properties and are good conductors of heat. Inorganic nanotubes are also being produced (electronicsnode.blogspot.com).

Their name nanotube originates from their size, because the diameter is only a few nanometers (about 50,000 times smaller than the thickness of a human hair), whereas the length can reach up to several millimeters (epdf.tips).

They can be single-walled carbon nanotube (SWCNT) having a diameter of less than 1 nm or multiwalled carbon nanotube (MWCNT), containing a number of concentrically interlinked nanotubes, having diameters of more than 100 nm. Their length can be several micrometers or even reach millimeters (www.nanowerk.com).

Fig. A.3 shows a multiwalled carbon nanotube (He et al., 2013).

Like graphene (which is their building block), "CNTs are chemically bonded with sp^2 bonds which is a very strong form of molecular interaction. This feature combined with carbon nanotubes' natural inclination to rope together via van der Waals forces provides the chance to develop ultrahigh strength and low-weight materials possessing highly conductive electrical and thermal properties. This makes them highly attractive for numerous applications.

The rolling-up direction (rolling-up or *chiral* vector) of the graphene layers determines the electrical properties of the nanotubes. Chirality describes the angle of the nanotube's hexagonal carbon-atom lattice.

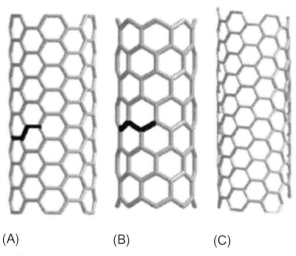

(A)   (B)   (C)

Fig. A.3 Carbon nanotube structures of armchair, zigzag, and chiral configurations. They differ in chiral angle and diameter: armchair carbon nanotubes share electrical properties similar to metals. The zigzag and chiral carbon nanotubes possess electrical properties similar to semiconductors. (A) Armchair. (B) Zigzag. (C) Chiral. *Reproduced with permission He H, Pham-Huy LA, Dramou P, Xiao D, Zuo P, Pham-Huy C. Carbon nanotubes: applications in pharmacy and medicine. Biomed Res Int 2013;2013:578290.*

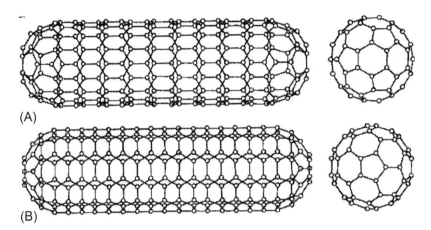

Fig. A.4 Idealized models of (A) zigzag and (B) armchair monolayer nanotubes. *Reproduced with permission from Zaporotskovaa IV, Borozninaa NP, Parkhomenkob YN, Lev V and Kozhitovb LV. Carbon nanotubes: sensor properties. A review, Mod Electron Mater 2016;2:95–105.*

"It was the so-called *Armchair* nanotubes because of the armchair-like shape of their edges. They have identical chiral indices and are highly preferred for their ideal conductivity. They are unlike *zigzag* nanotubes, which may be semiconductors. Turning a graphene sheet a mere 30 degrees will change the nanotube it forms from armchair to zigzag or inversely" (www.nanowerk.com). Fig. A.4 shows idealized models of (a) zigzag and (b) armchair monolayer nanotubes (Zaporotskovaa et al., 2016).

MWCNTs are always conducting and obtain the same level of conductivity like metals, whereas conductivity of SWCNT is dependent upon their chiral vector. They behave like a metal and are electrically conducting and show the properties of a semiconductor or nonconducting. For instance, a minor alteration in the pitch of the helicity can convert the tube from a metal into a large-gap semiconductor.

(www.nanowerk.com)

CNTs also have excellent mechanical and thermal properties, which make them interesting for developing new materials apart from their electrical properties, which they take over from graphene (Tables A.1 and A.2).

"These properties make CNTs best candidates for following applications:
- electronic devices
- chemical/electrochemical and biosensors
- transistors
- electron field emitters
- lithium-ion batteries
- white light sources
- hydrogen storage cells

Table A.1 Properties of CNTs.

Very light weight. Density is one-sixth of that of steel
Tensile strength is 400 times that of steel
Thermal conductivity is superior than that of diamond
Very high aspect ratio larger than 1000. They are extremely thin in relation to their length
A tip surface area near the theoretical limit (the smaller the tip surface area, the more concentrated the electric field, and higher the field improvement factor)
Highly chemically stable just like graphite and resist nearly any chemical impact unless they are simultaneously exposed to high temperatures and oxygen—a property that makes them extremely resistant to corrosion
Their hollow interior can be filled with various nanomaterials, separating and shielding them from the surrounding environment—a property that is very useful for nanomedicine applications such as drug delivery

https://www.nanowerk.com/introduction_to_nanotechnology_22.

Table A.2 Tabular representation of CNT properties.

Property (unit)	SWCNT	MWCNT
Diameter (nm)	0.4–3	1.4–100
Aspect ratio	4300	1250–3750
Specific surface area ($m^2 g^{-1}$)	150–790	50–850
Tensile strength (GPa)	22 ± 2	11–63
Young's modulus (TPa)	0.79–3.6	0.27–2.4
Thermal conductivity ($W\,m^{-1}\,K$)	3500–6600	600–6000
Band gap (eV)	0.1–2.2 (direct)	0–0.08 (direct)
Carrier mobility ($cm^2 V^{-1} s^{-1}$)	20–104	
Resistivity (Ω-cm)	10–6	
Current density ($A\,cm^{-2}$)	107–108	106–1010
Cost (per gram in USD)	50–400	0.1–25

Based on Venkataraman A, Amadi EU, Chen Y and Papadopoulos C. Carbon nanotube assembly and integration for applications Nanoscale Res Lett 2019;14:220. doi: 10.1186/s11671-019-3046-3; Peigney A, Laurent C, Flahaut E, Bacsa RR and Rousset A. Specific surface area of carbon nanotubes and bundles of carbon nanotubes. Carbon 2001;39:507–514; Wang X, Li Q, Xie J, Jin Z, Wang J, Li Y, Jiang K and Fan S. Fabrication of ultralong and electrically uniform single-walled carbon nanotubes on clean substrates. Nano Lett 2009;9:3137–3141; Li F, Cheng HM, Bai S, Su G and Dresselhaus MS. Tensile strength of single-walled carbon nanotubes directly measured from their macroscopic ropes. Appl Phys Lett 2000;77:3161–3163; Demczyk B, Wang Y, Cumings J, Hetman M, Han W, Zettl A and Ritchie. Direct mechanical measurement of the tensile strength and elastic modulus of multiwalled carbon nanotubes. Mater Sci Eng A 2002;334:173–178; Hone J, Llaguno MC, Biercuk MJ, Johnson AT, Batlogg B, Benes Z and Fischer JE. Thermal properties of carbon nanotubes and nanotube-based materials. Appl Phys A Mater Sci Process 2002;74:339–343; Qian D, Wagner GJ, Liu WK and Ruoff RS. Mechanics of carbon nanotubes. Appl Mech Rev 2002;55:495–533; Li JQ, Zhang Q, Chen G, Yoon SF, Ahn J, Wang SG, Zhou Q and Wang Q. Thermal conductivity of multiwalled carbon nanotubes. Phys Rev B 2002a;66(16):165440, Li Z, Chen J, Zhang X, Li Y and Fung K. Catalytic synthesized carbon nanostructures from methane using nanocrystalline Ni. Carbon 2002b;40:409–415; Dresselhaus MS, Dresselhaus G and Eklund PC. The science of fullerenes and carbon nanotubes: their properties and applications. Elsevier; 1996a, 20-Mar-1996—Science—965 pages. ISBN 0-12221-820-5, Dresselhaus MS, Dresselhaus G and, Eklund PC. Science of fullerenes and carbon nanotubes. San Diego: Academic; 1996b; Durkop T, Getty SA, Cobas E and Fuhrer MS. Extraordinary mobility in semiconducting carbon nanotubes. Nano Lett 2004;4(1):35–39; Lundstrom M, Wang Q, Javey A, et al. Ballistic carbon nanotube field effect transistors. Nature 2003;424:654–657. Wei BQ, Vajtai, R and Ajayan, PM. Reliability and current carrying capacity of carbon nanotubes. Appl Phys Lett 2001;79:1172–1174; Subramaniam C, Yamada T, Kobashi K,

Sekiguchi A, Futaba DN, Yumura M and Hata K. One hundred fold increase in current carrying capacity in a carbon nanotube-copper composite. Nat Commun 2013;4:2202–9; Wilder JW, Dekker C, Venema LC, Rinzler AG and Smalley RE. Electronic structure of atomically resolved carbon nanotubes. Nature 1998;391:59–62;
Wan R, Peng J, Zhang X and Len C. Band gaps and radii of metallic zigzag single wall carbon nanotubes. Phys B Condens Matter 2013;417:1–3; Baughman RH, Zakhidov AA and Heer WAD. Carbon nanotubes–the route toward applications. Science 2002;297:787–792; Al-Rub RKA, Ashour AI and Tyson BM. On the aspect ratio effect of multiwalled carbon nanotube reinforcements on the mechanical properties of cementitious nanocomposites. Construct Build Mater 2012;35:647–655; Schönenberger C, Bachtold A, Strunk C, Salvetat JP and Forr L. Interference and interaction in multi-wall carbon nanotubes. Appl Phys A Mater Sci Process 1999;69:283–295; Purewal, M, Hong, BH, Ravi, A, Chandra, B, Hone, J and Kim, P. Scaling of resistance and electron mean free path of single-walled carbon nanotubes. Phys Rev Lett 2007;98:186808; Lekawa-Raus A, Patmore J, Kurzepa L, Bulmer J and, Koziol K. Electrical properties of carbon nanotube based fibers and their future use in electrical wiring. Adv Funct Mater 2014;24:3661–3682; Zhang Q, Huang JQ, Qian WZ, Zhang YY and Wei F. The road for nanomaterials industry: a review of carbon nanotube production, post-treatment, and bulk applications for composites and energy storage. Small 2013;9:1237–1265; Chen W, Liu Y, Liu P, Xu C, Liu Y and Wang Q. A temperature-induced conductive coating via layer-by-layer assembly of functionalized graphene oxide and carbon nanotubes for a flexible, adjustable response time flame sensor. Chem Eng J 2018;353:115–125.

- cathode ray tubes
- electrostatic discharge
- electrical shielding" (www.nanowerk.com)

It should be noted that CNTs differ from carbon nanofibers (CNFs). Diameter of CNFs is about 200 nm, and the length is normally several micrometers long. Carbon fibers have been used for long time for strengthening compound, but they do not possess the same lattice structure as CNTs. Alternatively, they contain a blend of numerous forms of carbon and/or numerous layers of graphite, which are heaped at different angles on amorphous carbon (where atoms do not arrange themselves in organized structures). The properties of CNFs are similar to CNTs but have lower tensile strength because of their inconsistent structure. They are not hollow inside.

A.1 Discovery of carbon nanotubes

Several papers are being published every year on CNTs or allied areas. Sumio Iijima discovered CNTs in 1991. He published a paper in Nature on helical microtubules of graphitic carbon. This paper reported the discovery of **MWCNT**.

On taking a quick look at the scientific literature, one might get the impression that Iijima is the de facto discoverer of carbon nanotubes. Certainly, there is no disbelief that he has made two seminal contributions to the field, but an analysis of the literature suggests that certainly he is not the first one who has reported about CNTs.

A.2 Techniques for fabrication of CNTS

Following techniques are presently available for fabrication of CNTs:
- arc discharge
- laser ablation of graphite
- chemical vapor deposition (CVD)

In the arc discharge and laser ablation of graphite, combustion is performed electrically or by means of a laser. CNTs are developed in the gaseous phase, which are separated. All these methods need metals as catalysts. Examples of metal catalysts are iron, cobalt, and nickel.

A.2.1 Arc-discharge method

Iijima (1991) produced CNTs using the arc-discharge method from carbon soot of graphite electrode. This method uses a temperature of more than 1700°C for synthesizing CNTs (Ando et al., 2004). Two graphite electrodes, an anode and cathode, are used. These electrodes have diameter of 6 and 9 mm that are placed about 1 mm apart in a huge metal reactor (Fig. A.5).

"Inert gas is maintained at a constant high pressure inside the metal reactor. A direct current of ~100 A is passed with a potential difference of ~18 V. When the two

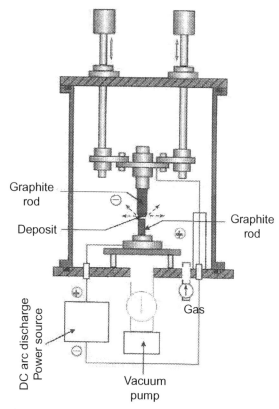

Fig. A.5 Carbon nanotubes arc discharge production method. *Reproduced with permission Purohit R, Purohit K, Rana S, Rana R.S and Patel, V. Carbon nanotubes and their growth methods. Procedia Mater Sci 2014;6: 716–728. https://doi.org/10.1016/j.mspro.2014.07.088.*

electrodes are brought closer, a discharge takes place leading to the formation of plasma. A carbonaceous deposit containing nanotubes is formed on the larger electrode. MWCNTs in the form of carbon soot of 1–3 nm inner diameter and ~2–25 nm outer diameter were deposited in the negative electrode. By doping the anode with metal catalysts such as cobalt, iron, or nickel and using pure graphite electrode as the cathode, SWCNTs could be grown up to a diameter of about 2–7 nm" (nanoscalereslett.springeropen.com; Ajayan and Ebbesen, 1992; Ma et al., 2010; Tanaka and Iijima, 2014; Iijima, 1991). This method can be used for growing large amounts of SWCNTs and MWCNTs. But the main downside of this method is that the yield is limited because of the use of metal catalysts, which introduces undesirable postreaction products needing purification (Venkataraman et al., 2019).

A.2.2 Laser ablation

This method is akin to the arc-discharge method, but it uses a continuous laser beam or a pulsed laser instead of arc-discharge (Henley et al., 2012). Fig. A.6 shows the schematic synthesis apparatus: (a) classical laser ablation technique and (b) ultrafast laser evaporation (free electron laser (FEL)) (Szabo et al., 2010).

The laser beam vaporizes a large graphite target in the presence of an inert gas such as helium, argon, and nitrogen in a quartz tube furnace at a temperature of ~1200°C. The vaporized carbon is then condensed, and CNTs get self-assembled on the cooler surface of the reactor (Guo et al., 1995a,b; Thess et al., 1996; Hafner et al., 1998). "MWCNTs are produced having an inner diameter of ~1–2 nm and an outer diameter of about 10 nm if both electrodes are made up of pure graphite. When the graphite target is doped with cobalt, iron, or nickel, the resultant deposit is rich in SWCNT bundles. The yield and quality of CNTs depend on the growth environment, for example, laser properties, composition of catalyst, growth temperature, selection of gases, and pressure. This technique is costly because of the need for laser beams of high power. One benefit of this

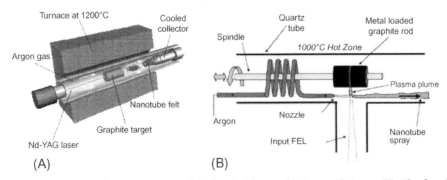

Fig. A.6 Schematic synthesis apparatus. (A) Classical laser ablation technique. (B) Ultrafast laser evaporation (free electron laser (FEL)). *Reproduced with permission from Szabo A, Perri C, Csato A, Giordano G, Vuon, D, Nagy, JB. Synthesis methods of carbon nanotubes and related materials. Materials 2010;3: 3092–3140.*

method is that post-growth purification is not as rigorous as in case of arc-discharge method because of lesser contamination" (nanoscalereslett.springeropen.com; Lebedkin et al., 2002).

A.2.3 CVD process

This process looks promising. Larger amount of CNTs are produced at reduced cost under more easily convenient conditions. In this process a metal catalyst (like iron) can be combined with carbon containing reaction gases (like hydrogen or carbon monoxide) to produce CNTs on the catalyst inside a furnace at high temperature. Fig. A.7 shows catalyzed chemical vapor deposition.

The CVD process can be performed with pure catalyst or can be supported by plasma. The latter process needs somewhat reduced temperatures (200–500°C) as compared with the catalytic process (up to 750°C) and produces "lawn-like" CNT growth.

"Although synthetic methods have been improved for obtaining high-purity carbon nanotubes, the formation of byproducts containing impurities such as metal encapsulated nanoparticles, metal particles in the tip of a carbon nanotube, and amorphous carbon has been a necessary phenomenon, as the metal nanoparticles are important for the nanotube growth.

These foreign nanoparticles and also the structural defects that takes place during synthesis have the adverse implication that the physicochemical properties of the produced CNTs are modified" (www.nanowerk.com).

That is the reason why CNTs should be purified using different techniques such as acid treatment or performing ultrasound after the completion of the process.

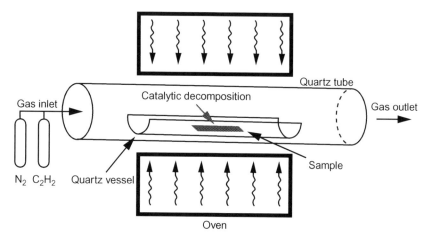

Fig. A.7 Catalyzed chemical vapor deposition. *Reproduced with permission from Purohit, R, Purohit, K, Rana, S, Rana, R.S and Patel, V. Carbon nanotubes and their growth methods. Procedia Mater Sci 2014;6:716–728. https://doi.org/10.1016/j.mspro.2014.07.088.*

A.3 Carbon nanotube uses and applications

CNTs are appropriate for nearly any application needing electrical and thermal conductivity, high strength, durability, and lightweight properties in comparison with traditional materials.

At present, CNTs are mostly used as additives to synthetics. On a commercial scale, CNTs are available in the form of powder in a highly agglomerated form. For CNTs to unfurl their properties, they should be untangled and spread uniformly on the substrate.

> Another necessity is that CNTs should be chemically linked with the substrate, example a plastic material. For that reason, CNTs are functionalized. Their surface is chemically adapted for optimal addition into different materials and for the precise application in question.
>
> CNTs can also be spun into fibers. This promises exciting possibilities for specialty textiles and can also help to realize a utopian project—the space elevator.
>
> **(www.nanowerk.com).**

A.3.1 Materials

CNT-enabled nanocomposites are receiving a great deal of consideration as a highly attractive option to traditional composite materials because of their mechanical, electrical, thermal, barrier, and chemical properties, for example, electrical conductivity, higher tensile strength, better heat distortion temperature, or flame retardancy.

These materials are offering improved wear resistance and breaking strength, antistatic properties, and also reduction in weight. For example, advanced CNT composites can reduce the weight of aircraft and spacecraft by up to 30%. These composite materials have already found use in the following areas:

- Sporting goods
- Yachting
- Textiles
- Automotive, aeronautics, and space
- Industrial engineering
- Electrostatic charge protection and radiation shielding with CNT-based nanofoams and aerogels

A.3.2 Catalysis

CNTs are drawing attention for catalysis because of their remarkably high surface area and their capability to get attached to any chemical species to their sidewalls. CNTs are used as catalysts in several important chemical processes, but it is not easy to control their catalytic activity.

Primarily, CNTs are combined with molecules through covalent bonds, which are very strong. This leads to very stable compounds. Such connection entails a change in the nanotube structure and so in its properties.

It would be analogous to nailing an advertisement to a post using a thumbtack: the union is strong, but it leaves a hole in both the advertisement and the post. Weak noncovalent forces have also been explored, which keep nanotubes intact, but normally yield kinetically unstable compounds. The comparison in this case would be to tape the advert to the post. The advertisement or the post is not damaged, but the union is much feeble. For solving this problem, scientists are developing methods for the chemical modification of CNTs by mechanical bonding, the first example of mechanically interlocked carbon nanotubes (MINTs). These are stable like covalent compounds but at the same time as respectful of the initial structure as the noncovalent compounds.

(www.nanowerk.com)

A.3.3 Transistors

In spite of the increase of graphene and other two-dimensional materials, SWCNTs, which are semiconducting, are considered as good candidates not only for the next generation of high-performance, ultrascaled, and thin-film transistors but also for optoelectronic devices for replacing silicon electronics.

One of the main questions is if CNT transistors can provide performance benefits over silicon at sub-10 nm lengths. There have been mixed views in the nanoelectronics community concerning whether or not CNT transistors would maintain their remarkable performance at enormously scaled lengths. Some have the opinion that the very small mass of the carriers may contribute to a tunneling phenomenon causing the devices to collapse to approximately 15 nm. This view is supported by the few studies that examined nanotube devices at such dimensions.

For the present, others get convinced that the ultrathin body of SWCNTs—only 1 nm in diameter—may allow for superb transistor behavior even to the sub-10 nm level.

So far, promising experimental results have been achieved, and at the moment, there remain several challenges concerning incorporating CNT transistors into chip manufacturing on an industrial scale.

A.3.4 Sensors

The group of Cees Dekker paved the way for the development of CNT-based electrochemical nanosensors by demonstrating the possibilities of SWCNTs as quantum wires and their effectiveness in the development of field-effect transistors.

Although CNTs are strong, their electrical properties are very susceptible to the effects of charge transfer and chemical doping by several molecules. Most sensors based on CNTs are field-effect transistors (FETs). CNTs–FETs have been extensively used for identifying gases, for instance, greenhouse gases in environmental applications.

The functionalization of CNTs is essential to make them selective to the target analyte. Diverse types of sensors are based on molecular recognition interactions between functionalism CNT and target analytes.

(www.nanowerk.com)

For example, flexible hydrogen sensors using SWCNTs decorated with palladium nanoparticles have been developed.

Ink formulations based on CNT dispersions are useful for printed electronic applications: thin-film transistors, light-emitting devices transparent electrodes, RFID tags, and solar cells.

A.3.5 Electrodes

CNTs are used as electrodes for biological and chemical sensing applications and several other electrochemical studies. With their exceptional one-dimensional molecular geometry of a large surface area coupled with their outstanding electrical properties, CNTs are becoming essential for molecular engineering the electrode surfaces where developing the electrochemical devices with region-specific electron-transfer potential is of vital significance.

A.3.6 Displays

Because of their high electrical conductivity and the unbelievable sharpness of their tip, CNTs are the most promising material for field emitters. Examples are CNTs as electron emitters for field emission displays (FED). "The smaller the tips' radius of curvature, the more concentrated the electric field, the higher field emission."

FED technology is making possible a new class of large area, high-resolution, reduced cost flat panel displays. But FED manufacturing needs CNT to be grown in exact sizes and densities. The voltage is influenced by height, diameter, and tip sharpness, whereas current is affected by density.

A.3.7 Buckypapers

There are various applications of buckypapers. It is thermally conductive material and may lead to the development of more capable heat sinks for chips; a more energy-efficient and lighter background illumination material for displays; a shielding material for electronic circuits from electromagnetic interference because of its abnormally high current-carrying capacity.

A.3.8 Optoelectronic and photonic applications

Whereas individual nanotubes produce distinct fine peaks in optical absorption and emission, macroscopic structures containing various CNTs assembled together also show remarkable optical performance.

For instance, a millimeter-long bundle of aligned MWCNTs releases polarized glowing light by heating of electrical current, and SWCNT bundles give higher brightness emission at reduced voltage in comparison with traditional tungsten filaments.

A.3.9 Nanomedicine and biotechnology

CNTs or graphene is of immense interest to biomedical engineers working on nanotechnology applications. The physical properties of CNTs are of great value for manufacturing advanced biomaterials.

By chemical modification, specific moieties, for example, functional groups, molecules, and polymers, can be introduced in CNTs for imparting properties suitable for biological applications, for instance, increased solubility and biocompatibility, increased material compatibility, and cellular responsiveness.

Nitrogen-doped CNTs have been developed for drug delivery applications.

But the matter of cytotoxicity of CNTs is an area that is attracting a lot of interest and has not yet resulted in any clear-cut answer. Therefore more organized biological assessments of CNTs having several chemical and physical properties are required for determining their exact pharmacokinetics, cytotoxicity, and optimal dosages.

A.3.10 Filtration

Membranes having high flow are becoming an important part of future high energy-efficient water purification systems. Efficient water transport demonstrated in CNTs with openings of less than 1 nm has been demonstrated.

When inserted in fatty membranes, the nanotubes compress the entering water molecules into a single file chain, which results in extremely fast transport. The flow was 10 times quicker than in wider CNTs and 6 times quicker than in the best biological membrane.

CNTs also have been used to demonstrate protective textiles with ultrabreathable membranes. These membranes provide rates of water vapor transport, which exceed those of commercial breathable fabrics like Gore-Tex, even though the CNT pores are only a few nanometers wide.

(www.nanowerk.com)

Importantly, CNTs also provide protection from biological agents because of their very small pore size that is lesser than 5 nm wide. Biological threats like bacteria or viruses are much larger and usually more than 10 nm in size.

For making these membranes also protect from chemical agents, which are much smaller in size, CNT surfaces have been modified with chemical threat-responsive functional groups. These groups will sense and block the threat like gatekeepers on the pore entrance.

(www.nanowerk.com)

A.4 Factors affecting the growth of carbon nanotubes

Parameters affecting the growth of CNTs are the catalyst, carrier gas, substrate, reaction time, temperature, and the carrier gas flow rate. Most effective catalysts are iron, cobalt, and nickel. The justification for using these metals lies in the phase diagrams for these metals and carbons. Carbon has finite solubility in these metals at a high temperature. This results in the formation of metal-carbon solutions and therefore the aforesaid growth mechanism. In the growth process, catalysts play a vital role to control the structure of SWCNTs.

> *The size-controlled synthesis of iron-based nanoparticles and the CVD growth of SWCNTs, particularly the horizontally aligned SWCNTs, catalyzed by these produced nanoparticles was studied. Some new catalysts were also produced. Among them, copper is found to be a better catalyst for growing SWCNT arrays on both silicon and quartz substrates, and lead was a unique catalyst from which one can obtain SWCNTs without any metallic contaminant. SWCNTs produced with both copper and lead was very suitable for building high-performance nanodevices. These studies are very helpful for further understanding the growth mechanism of SWCNTs. The growth rate of CNT is a function of catalyst particle size and the diffusion rate of carbon through the catalyst. As the catalyst particle size increases, the growth rate decreases. The growth rate of CNTs is directly proportional to the diffusion rate of carbon through the catalyst.*
>
> **(Mubarak et al., 2014; Kim et al. 2003)**

Extensively used catalysts in CNT synthesis are cobalt, iron, titanium, nickel, a few zeolites and combinations of these metals, and oxides (Nagaraju et al., 2002; Sinha et al., 2000; Couteau et al., 2003; Rohmund et al., 2000, Chen et al., 2002, 2003; Klinke et al., 2001; Zeng et al., 2002; Mukhopadhyay et al., 1999; Wang et al., 2001; Ning et al., 2002; Su et al., 2000; Han et al., 2002). Another factor affecting the productivity of the catalyst is the dispersion of the catalyst. Large particles and aggregates instead of fine and well-dispersed particles may be inactive for nanotube growth, and well-dispersed catalyst systems may result in narrower size distribution of the synthesized nanotubes (Kong et al., 1998; Hernadi et al., 1996). Ethylene, acetylene, methane, benzene, xylene, and carbon monoxide are generally used CNT precursors (Kong et al., 1998; Fan et al., 1999; Li et al., 1996; Douven et al., 2012; Sen et al., 1997; Satishkumar et al., 1998; Nikolaev et al., 1999; Nikolaev, 2004; Wei et al., 2002; Endo et al., 1991).

The CNT growth from pyrolysis of benzene at **1100**°C was studied by Endo et al. (1992, 1993). Clear helical MWCNTs at 700°C from acetylene were obtained by Jose-Yacaman et al. (1993). The carrier gases—hydrogen and argon—also affect the growth of CNTs. Hydrogen provides a reducing atmosphere and scavenges oxygen, so it is the most efficient carrier gas.

But high amount of hydrogen will drive the reaction in the opposite direction, which is not wanted. Nitrogen is effective for the growth of bamboo-type CNTs by retarding the passivation of the catalyst and increase carbon diffusion through the catalyst. Ammonia actually provides large quantity of hydrogen, which represses the carbon supply and increases the formation of tube. The growth of CNTs strongly depends on substrate. The high surface roughness pores and substrate defects encourage the growth of nanotubes having surface defects. The substrate crystallinity also affects the growth of CNTs and nucleation (Mubarak et al., 2014).

The growth of SWCNTs is favored on crystalline surface of substrate, whereas MWCNT's growth is favored on amorphous and polycrystalline surface. The mechanism of CNT group can be grouped into two modes depending on the interaction of the catalyst with the substrate. These are base growth mode and the tips growth mode. The reaction rate is a strong function of temperature; therefore the concentration of each species varies with temperature. At reduced temperature a reduction in the formation of CNT formation was seen perhaps because of the limited deactivation of catalyst (Cabero et al., 2004).

Lower synthesis temperatures result in reduced CNT yield in the product (Sinha et al., 2000). Reaction temperature also plays a significant role in the alignment properties and diameter of the synthesized nanotubes (Singh et al., 2003). Usually the CNT growth temperature used is between 550°C and 1000°C, and reaction temperature varies according to the catalyst support material pair.

Zhu et al. (2003) used thermal CVD for synthesizing the CNT on graphite fibers. It was found that the "CNTs can only be grown in a limited temperature range. At reduced temperatures, only a carbon layer gets formed on the fiber surface. But, at high temperatures, the diffusion of iron particles into carbon fibers improved, and the growth reduced" (studentsrepo.um.edu.my).

Li et al. (2002b) produced the CNTs with nanocrystalline Ni/metallic oxide catalyst from decomposition of methane. When the reaction temperature was higher than 1000 K, the CNTs were obtained. At lower temperatures the formation of carbon fibers was seen.

Production of CNTs below a temperature of 400°C was studied. "No significant change was noted in the particle size distribution for nickel particles, showing that the temperature is too low for nucleation and growth. The typical reaction time is about 60 min, but the reaction time also depends on the CNTs amount. Moreover, deviations of the optimum reaction time for the specific synthesis suffer from the final product quality" (Mubarak et al., 2014; Chiang and Sankaran, 2008; Nagaraju et al., 2002).

The effect of flow rate of precursor and carrier gas was examined by Han et al. (2002).

Increasing the flow rate of carrier gas and keeping the precursor flow rate constant, the mean diameter of CNTs decreased. If the flow rates of carrier and precursor gas are increased, the diameter remains constant, but the growth rate increases.

(studentsrepo.um.edu.my)

The decomposition of methane for MWCNTs growth among other factors such as operating conditions is dependent upon catalyst and the loading of the active species for methane decomposition, catalyst preparation, and pretreatment such as the calcination and reduction in hydrogen. A kinetic model is developed for the decomposition of methane during growth of MWCNTs over 20 wt% nickel-alumina catalyst by CVD. The activation energy for methane decomposition on Ni-Al_2O_3 catalyst was 37.1 ± 3.5 kJ/mol.

(Mubarak et al., 2014; Inusa et al. 2012)

Mechanism of growth and mass production of CNTs by CVD with the effect of material aspect, for example, roles of hydrocarbon, catalyst, and catalyst support, was studied (Mukul and Yoshinor, 2010). Growth-control aspects, for instance, the effects of temperature vapor pressure and catalyst concentration on CNT diameter distribution and single- or multiwalled CNTs, were discussed.

Out of different methods, catalytic CVD in a fluidized bed is a good method for production of CNTs on a large scale. Bulk production of CNTs depends on various parameters. These are flow rate of gas, hydrocarbon concentration, temperature, size and shape of the initial agglomerate, and catalyst loading and size. Flow rate of gas has an effect on the mass transfer coefficient. The size and shape of the agglomerate affect the internal diffusion of gas within pore. The reaction rate is determined by the temperature, hydrocarbon concentration, and catalyst dose. The morphology of CNTs is dependent upon the type and size of catalyst and the carbon supply rate (Danafar et al., 2009).

A.5 Purification of CNTs

In CNTs a large amount of impurities such as metal particles, amorphous carbon, and multishell are present. For purification of CNTs, different steps are involved. Purification involves separating nanotubes from nonnanotube impurities included in the raw products or from nanotubes having undesired numbers of walls. Purification has been an important effort since the discovery of carbon nanotubes (www.intechopen.com). Several papers have been published discussing purification process (Park et al., 2006; Haddon et al., 2004; Farkas et al., 2002).

The industrial methods use oxidation and acid-refluxing methods, which affect the structure of tubes. Purification difficulties are great due to the limitation of liquid chromatography and insolubility of CNT.

Purification removes amorphous carbon from CNTs, improves surface area, decomposes functional groups obstructing the entrance of the pores, or simulates additional functional groups.

Most of these methods are combined with each other for improving the purification and removing different impurities at the same time. These methods are as follows (www.intechopen.com):
- oxidation
- acid treatment
- annealing and thermal treatment
- ultrasonication
- microfiltration

A.5.1 Oxidation

Oxidation removes CNT impurities. The impurities and CNTs are oxidized. However, the harm to CNTs is less in comparison with the harm to the impurities. This method is favored with regard to the impurities that are generally metal catalysts acting as oxidizing catalysts (Hajime et al., 2002; Borowiak-Palen and Pichler, 2002).

On the whole the efficiency and yield depend upon several factors (Borowiak-Palen and Pichler, 2002). These are listed in the succeeding text:
- metal content
- oxidation time
- environment
- oxidizing agent
- temperature

A.5.2 Acid treatment

When the sample is refluxed with acid, the amount of metal particles and amorphous carbon is reduced. Different types of acids are used. These are hydrochloric acid, nitric acid, and sulfuric acid. The ideal refluxing acid is hydrochloric acid. With nitric acid the acid affected the metal catalyst only, and no effects was seen on the CNTs and the other carbon particles (Hajime et al., 2002; Borowiak-Palen and Pichler, 2002; Kajiura et al., 2002).

Fig. A.8 shows the SEM images of CNTs after and before purification with hydrochloric acid (Jahanshahi and Seresht, 2009a,b). If a treatment in hydrochloric acid is used, the acid has a slight effect on the CNTs and other carbon particles (Kajiura et al., 2002; Chiang et al., 2001).

A.5.3 Annealing and thermal treatment

High temperature affects the production and paralyzes the graphitic carbon and the short fullerenes. At high temperature the metal is melted and can be removed (Kajiura et al., 2002).

A.5.4 Ultrasonication

This method is based on the separation of particles because of ultrasonic vibrations. The agglomerates of different nanoparticles are more dispersed by this technique. The separation of the particles is highly dependent on the surfactant, solvent, and the reagents used (Borowiak-Palen and Pichler, 2002; Kajiura et al., 2002; Chiang et al., 2001).

When acid is used, the CNT purity depends on the sonication time. During the tube's vibration to the acid for a short duration, only the metal is solvated. However, in a more extended period, the CNTs are also chemically cut (Kajiura et al., 2002).

A.5.5 Microfiltration

Microfiltration is based on particle size. Generally, CNTs and a small amount of carbon nanoparticles are trapped in a filter. The other nanoparticles (catalyst metal, fullerenes, and carbon nanoparticles) are passing through the filter. A special form of filtration is cross

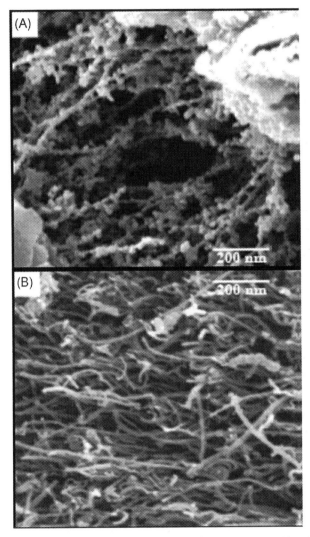

Fig. A.8 The SEM images of CNTs (A) after (B) before purification stages with HCl. *Reproduced with permission from Jahanshahi M, and Seresht R.J. Catalysts effects on the production of carbon nanotubes by an automatic arc discharge set up in solution. Phys Status Solidi C 2009a;6:2174–2178.*

flow filtration. Through a bore of fiber, the filtrate is pumped down at head pressure from a reservoir, and the major fraction of the fast flowing solution is reverted to the same reservoir in order to be cycled through the fiber again. A fast hydrodynamic flow down the fiber bore sweeps the membrane surface and avoids building up of a filter cake (fac.ksu.edu.sa; Haddon et al., 2004; Kajiura et al., 2002; Chiang et al., 2001; Moon and An, 2001; Borowiak-Palen and Pichler, 2002).

A.6 Morphological and structural characterizations

For investigating the morphological and structural characterizations of the CNTs, a number of methods can be used. It is essential to characterize and determine the quality and properties of the CNTs, as its applications need certification of properties and functions (Mawhinney et al., 2000).

But only few methods are able to characterize CNTs at the individual level like scanning tunneling microscopy (STM) and transmission electronic microscopy (TEM). X-ray photoelectron spectroscopy is needed for determining the chemical structure of CNTs in spite of the fact that Raman spectroscopy is frequently used as universal characterization method.

A.6.1 Electron microscopy (SEM and TEM)

The morphology, dimensions, and orientation of CNTs can be determined by using SEM and TEM techniques having high resolution (Chiang et al., 2001; Bandow and Rao, 1997; Moon and An, 2001; Jahanshahi and Kiadehi, 2013; Mawhinney et al., 2000; Li et al., 2001).

TEM is used for measurement of the outer and inner radius and linear electron absorption coefficient of CNTs (Gommes et al., 2003). This method is used to study CNTs before and after annealing and notices a substantial increase of the electron absorption coefficient. The intershell spacing of MWNTs was studied by Kiang and Endo (1998) using TEM.

Fig. A.9 shows SEM micrographs of the pristine nanotubes (PNTs) (a) and cut nanotubes (CNTs) (b) (Maurin et al., 2001). After the mechanical treatment the surface is clearly modified, and nanotubes with short lengths are seen.

Fig. A.10 shows typical TEM micrograph of multiwalled carbon nanotubes.

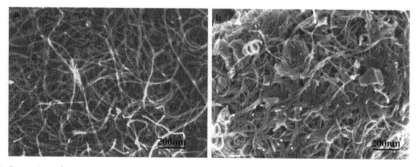

Fig. A.9 Scanning electron microscopy (SEM) micrographs of the pristine nanotubes (PNTs) (A) and cut nanotubes (CNTs) (B). After the mechanical treatment the general aspect of the surface is clearly modified, and nanotubes with short lengths are observed. *Reproduced with permission from Maurin G, Stepanek I, Bernier P, Colomer J-F, Nagy JB and Henn F. Segmented and opened multi-walled carbon nanotubes. Carbon 2001;39:1273–8.*

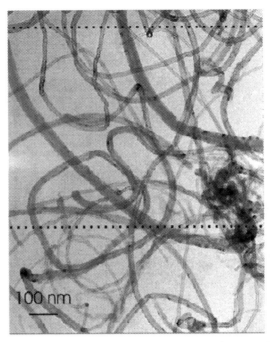

Fig. A.10 Typical transmission electron microscopy micrograph of multiwalled carbon nanotubes. *Reproduced with permission from Gommes C, Blacher S, Masenelli-Varlot K, Bossuot Ch, McRae E, Fonseca A, Nagy JB and Pirard JP. Image analysis characterization of multi-walled carbon nanotubes, Carbon 2003;41: 2561–2572.*

References

Abdalla S, Al-Marzouki F, Al-Ghamdi AA, Abdel-Daiem A. Different technical applications of carbon nanotubes. Nanoscale Res Lett 2015;10:1–10.
Ajayan PM. Nanotubes from carbon. Chem Rev 1999;99:1787–800.
Ajayan PM, Ebbesen TW. Large-scale synthesis of carbon nanotubes. Nature 1992;358:220–2.
Ando Y, Zhao X, Sugai T, Kumar M. Growing carbon nanotubes. Mater Today 2004;7:22–9.
Bandow S, Rao A. Purification of single-wall carbon nanotubes by micro-filtration. J Phys Chem B 1997;101:8839–42.
Bethune DS, Klang CH, de Vries MS, Gorman G, Savoy R, Vazquez J, Beyers R. Cobalt-catalysed growth of carbon nanotubes with single- atomic-layer walls. Nature 1993;363:605–7.
Borowiak-Palen E, Pichler T. Reduced diameter distribution of single-wall carbon nanotubes by selective oxidation. Chem Phys Lett 2002;363:567–72.
Cabero MP, Monzonb A, Ramosa IR, Ruiz A. Syntheses of CNTs over several iron-supported catalysts: influence of the metallic precursors. Catal Today 2004;93–95:681–7.
Chen J, Zhan X, Li Y, Fung K. Catalytic synthesized carbon nanostructures from methane using nanocrystalline Ni. Carbon 2002;40:409–15.
Chen M, Chen CM, Koo HS, Chen CF. Catalyzed growth model of CNTs by microwave plasma vapor deposition using CH_4 & CO_2 gas mixture Diamond Relat Mater 2003;12:1829–1838.

Chiang WH, Sankaran RM. Synergistic effects in bimetallic nanoparticles for low temperature carbon nanotube growth. Adv Mater 2008;20:4857–61.

Chiang IW, Brinson BE, Huang AY, Willis PA, Bronikowski MJ, Margrave JL, Smalley RE, Hauge RH. Purification and characterization of single-wall carbon nanotubes (SWNTs) obtained from the gas-phase decomposition of CO (HiPco process). J Phys Chem B 2001;105(35):8297–301.

Coleman JN, Khan U, Blau WJ, Gun'ko YK. Small but strong: a review of the mechanical properties of carbon nanotube–polymer composites. Carbon 2006;44:1624–52.

Colomer J, Henrard L, Lambin P, Van Tendeloo G. Electron diffraction and microscopy of single-wall carbon nanotube bundles produced by different methods. Eur Phys J B 2002;27:111–8 18. Chen W, Liu P, Liu Y.

Couteau E, Hernadi K, Seo JW, Nga LT, Miko C, Gaal R, Forro L. CVD synthesis of high-purity multi-walled carbon nanotubes using $CaCO_3$ catalyst support for large-scale production. Chem Phys Lett 2003;378(1–2):9–17.

Danafar F, Fakhru'l-Razi A, Salleh MAM, Biak DRA. Fluidized bed catalytic chemical vapor deposition synthesis of carbon nanotubes—A review. Chem Eng J 2009;155(1):37–48.

Demczyk B, Wang Y, Cumings J, Hetman M, Han W, Zettl A, Ritchie RO. Direct mechanical measurement of the tensile strength and elastic modulus of multiwalled carbon nanotubes. Mater Sci Eng A 2002;334:173–8.

Douven S, Pirard SL, Chan FY, Pirard R, Heyen G, Pirard JP. Large-scale synthesis of multi-walled carbon nanotubes in a continuous inclined mobile-bed rotating reactor by the catalytic chemical vapour deposition process using methane as carbon source. Chem Eng J 2012;188:113–25.

Dresselhaus MS, Dresselhaus G, Eklund PC. The science of fullerenes and carbon nanotubes: their properties and applications. Elsevier; 1996a 20-Mar-1996 - Science - 965 pages ISBN 0-12221-820-5.

Dresselhaus MS, Dresselhaus G, Eklund PC. Science of fullerenes and carbon nanotubes. San Diego: Academic; 1996b.

Dresselhaus MS, Dresselhaus G, Avouris P. Carbon nanotubes: synthesis, structure, properties, and applications. Springer-Verlag; 2000 (2000) ISBN 3-54041-086-4.

Dupuis AC. The catalyst in the CCVD of carbon nanotubes—a review. Prog Mater Sci 2005;50:929–61.

Ebbesen TW. Carbon nanotubes. Ann Rev Mater Sci 1994;24:235.

Ebbesen TW. Carbon nanotubes—preparation and properties. CRC Press; 1996 ISBN 0-84939-602-6.

Ebbesen TW, Ajayan PM. Large-scale synthesis of carbon nanotubes. Nature 1992;358:220–222.

Endo M, Fijiwara H and Fukunaga E (1991). 18th Meeting Japanese Carbon Society, Japanese Carbon Society, Saitama, December (1991), p. 34–35.

Endo M, Takeuchi K, Igarashi S, Kobori K, Shiraishi M and Kroto HW (1992). 19th Meeting Japanese Carbon Society, Japanese Carbon Society, Kyoto, December, p. 192.

Endo M, Takeuchi K, Igarashi S, Kobori K, Shiraishi M, Kroto HW. The production and structure of pyrolytic carbon nanotubes (PCNTs). J Phys Chem Solid 1993;54:1841–8.

Fan S, Chapline MG, Franklin NR, Tombler TW, Cassell AM, Dai H. Self-oriented regular arrays of carbon nanotubes and their field emission properties. Science 1999;283:512–4.

Farkas E, Anderson ME, Chen ZH, Rinzler AG. Length sorting cut single wall carbon nanotubes by high performance liquid chromatography. Chem Phys Lett 2002;363:111–6.

Ferrier M, De Martino A, Kasumov A, Gueron S, Kociak M, Egger R, Bouchiat H. Superconductivity in ropes of carbon nanotubes. Solid State Commun 2004;131:615–23.

Fischer J, Dai H, Thess A, Lee R, Hanjani NM, Dehaas DL, Smalley RE. Metallic resistivity in crystalline ropes of single-wall carbon nanotubes. Phys Rev B 1997;55:R4921–4.

Frank B, Rinaldi A, Blume R, Schlögl R, Su DS. Oxidation stability of multiwalled carbon nanotubes for catalytic applications. Chem Mater 2010;22:4462–70.

Gao B, Bower C, Lorentzen JD, Fleming L, Kleinhammes A, Tang XP, McNeil LE, Wu Y, Zhou O. Enhanced saturation lithium composition in ball-milled single-walled carbon nanotubes. Chem Phys Lett 2000;327:69–75.

Gommes C, Blacher S, Masenelli-Varlot K, Bossuot C, McRae E, Fonseca A, Nagy JB, Pirard JP. Image analysis characterization of multi-walled carbon nanotubes. Carbon 2003;41:2561–72.

Guo T, Nikolev P, Rinzler AG, Tomanek D, Daniel T, Colbert DT, Smalley RE. Self-assembly of tubular fullerenes. J Phys Chem 1995a;99:10694–7.

Guo T, Nikolaev P, Thess A, Colbert DT, Smalley RE. Catalytic growth of single-walled nanotubes by laser vaporization. Chem Phys Lett 1995b;243:49–54.

Haddon RC, Sippel J, Rinzler AG, Papadimitrakopoulos F. Purification and separation of carbon nanotubes. MRS Bull 2004;29:252–9.

Hafner JH, Bronikowski MJ, Azamian BR, Nikolaev P, Rinzler AG, Colbert DT, Smith K, Smalley RE. Catalytic growth of singlewall carbon nanotubes from metal particles. Chem Phys Lett 1998;296:195–202.

Hajime G, Terumi F, Yoshiya F, Toshiyuki O. Method of purifying single wall carbon nanotubes from metal catalyst impurities. Japan: Honda Giken Kogyo Kabushiki Kaisha; 2002.

Han JH, Choi SH, Lee TY, Yoo JB, Park CY, Jeong TW, Kim HJ, Park YJ, Han IT, Heo JN, Lee CS, Lee JH, Yu SG, Yi WK. Field emission properties of modified carbon nanotubes grown on Fe-coated glass using PECVD with carbon monoxide. Phys B Condens Matter 2002;323:182–3.

Harris PF. Carbon nanotubes and related structures: new materials for the twenty-first century. Cambridge University Press; 1999 ISBN 0-521-55446-2.

He H, Pham-Huy LA, Dramou P, Xiao D, Zuo P, Pham-Huy C. Carbon nanotubes: applications in pharmacy and medicine. Biomed Res Int 2013;2013:578290.

Henley S, Anguita J, Silva S. Synthesis of carbon nanotubes. Netherlands: Springer; 2012.

Hernadi K, Fonseca A, Nagy JB, Bernaerts D, Lucas A. Fe-catalyzed carbon nanotube formation. Carbon 1996;34:1249–57.

Hone J, Llaguno MC, Biercuk MJ, Johnson AT, Batlogg B, Benes Z, Fischer JE. Thermal properties of carbon nanotubes and nanotube-based materials. Appl Phys A Mater Sci Process 2002;74:339–43.

Iijima S. Helical microtubules of graphitic carbon. Nature 1991;354:56–8.

Iijima S, Ichihashi T. Single-shell carbon nanotubes of 1-nm diameter. Nature 1993;363:603.

Inusa A, Nataphan S, Jose EH. Diamond Relat Mater 2012;23(2012):76–82.

Jahanshahi M, Kiadehi AD. Fabrication, purification and characterization of carbon nanotubes: arc-discharge in liquid media (ADLM), In: Suzuki S, editor. Syntheses and applications of carbon nanotubes and their composites. IntechOpen; 2013. https://doi.org/10.5772/51116 Available from:https://www.intechopen.com/books/syntheses-and-applications-of-carbon-nanotubes-and-their-composites/fabrication-purification-and-characterization-of-carbon-nanotubes-arc-discharge-in-liquid-media-adlm.

Jahanshahi M, Seresht RJ. Catalysts effects on the production of carbon nanotubes by an automatic arc discharge set up in solution. Phys Status Solidi C 2009a;6:2174–8.

Jahanshahi M, Seresht RJ. Catalysts effects on the production of carbon nanotubes by an automatic arc discharge set up in solution. Phys Status Solidi C 2009b;6:2174–8 Wiley VCH.

Javey A, Guo J, Wang Q, Lundstrom M, Dai H. Ballistic carbon nanotube field effect transistors. Nature 2003;424:654–7.

Jose-Yacaman M, Miki-Yoshida M, Rendon L, Santiesteban JG. Catalytic growth of carbon microtubules with fullerene structure. Appl Phys Lett 1993;62(6):657–9.

Kajiura H, Tsutsui S, Huang HJ, Murakami Y. High-quality singlewalled purification from arc-produced soot. Chem Phys Lett 2002;364:586–92.

Kiang CH, Endo M. Size effect in carbon nanotubes. Phys Rev Lett 1998;81:1869–72.

Kim NS, Lee YT, Park J, Han JB, Choi YS, Choi SYC, Choo JC, Lee GH. Vertically aligned carbon nanotubes grown by pyrolysis of iron, cobalt, and nickel phthalocyanines. J Phys Chem B 2003;107(35):9249–55.

Klinke C, Bonard JM, Kern K. Comparative study of the catalytic growth of patterned carbon nanotube films. Surf Sci 2001;492(1):195–201.

Kong J, Cassell AM, Dai HJ. Chemical vapor deposition of methane for single-walled carbon nanotubes. Chem Phys Lett 1998;292:567–74.

Kroto HW, Fischer DCJE. The fullerenes: new horizons for the chemistry, physics and astrophysics of carbon. Cambridge: Cambridge University Press; 1993.

Kroto HW, Heath JR, O'Brien SC, Curl RF, Smalley REC. C60: buckminsterfullerene. Nature 1985;318:162–3.

Kumar M, Ando Y. Chemical vapor deposition of carbon nanotubes: a review on growth mechanism and mass production. J Nanosci Nanotechnol 2010;10(6):3739–58.

Lebedkin S, Schweiss P, Renker B, Hennrich F, Neumaier M, Stoermer C, Kappes MM. Single-wall carbon nanotubes with diameters approaching 6 nm obtained by laser vaporization. Carbon 2002;40:417–23.

Lee MW, Haniff MASM, Teh AS, Bien DC, Chen SK. Effect of Co and Ni nanoparticles formation on carbon nanotubes growth via PECVD. J Exp Nanosci 2015;10:1232–41.

Li WZ, Xie SS, Qian LX, Chang BH, Zou BS, Zhou WY, Zhao RA, Wang G. Large-scale synthesis of aligned carbon nanotubes. Science 1996;274:1701–3.

Li F, Cheng HM, Bai S, Su G, Dresselhaus MS. Tensile strength of single-walled carbon nanotubes directly measured from their macroscopic ropes. Appl Phys Lett 2000;77:3161–3.

Li W, Wen J, Tu Y, Ren Z. Effect of gas pressure on the growth and structure of carbon nanotubes by chemical vapor deposition. Appl Phys A 2001;73:259–64.

Li JQ, Zhang Q, Chen G, Yoon SF, Ahn J, Wang SG, Zhou Q, Wang Q. Thermal conductivity of multi-walled carbon nanotubes. Phys Rev B 2002a;66(16):165440.

Li Z, Chen J, Zhang X, Li Y, Fung K. Catalytic synthesized carbon nanostructures from methane using nanocrystalline Ni. Carbon 2002b;40:409–15.

Li Y, Cui RL, Ding L, Liu Y, Zhou WW, Zhang Y. How catalysts affect the growth of single-walled carbon nanotubes on substrates. Adv Mater 2010;22:1508–15.

Lundstrom M, Wang Q, Javey A. Ballistic carbon nanotube field effect transistors. Nature 2003;424:654–7.

Ma J, Wang JN, Tsai C, Nussinov R, Buyong MA. Diameters of single-walled carbon nanotubes (SWCNTs) and related nanochemistry and nanobiology. Front Mater Sci China 2010;4:17–28.

Martel R, Schmidt T, Shea HR, Hertel T, Avouris P. Single- and multi-wall carbon nanotube field-effect transistors. Appl Phys Lett 1998;73:2447–9.

Maurin G, Stepanek I, Bernier P, Colomer J-F, Nagy JB, Henn F. Segmented and opened multi-walled carbon nanotubes. Carbon 2001;39:1273–8.

Mawhinney DB, Naumenko V, Kuznetsova A, Yates JT, Liu J, Smalley RE. Surface defect site density on single walled carbon nanotubes by titration. Chem Phys Lett 2000;324:213–6.

Meyyappan M. Carbon nanotubes: science and applications. CRC Press; 2004 ISBN 0-84932-111-5.

Moon JM, An KH. High-yield purification process of singlewalled carbon nanotubes. J Phys Chem B 2001;105:5677–81.

Mubarak NM, Abdullah EC, Jayakumar NS, Sahu JN. An overview on methods for the production of carbon nanotubes. J Ind Eng Chem 2014;20:1186–97.

Mukhopadhyay K, Koshio A, Sugai T, Tanaka N, Shinohara H, Konya Z, Nagy JB. Bulk production of quasi-aligned carbon nanotube bundles by the catalytic chemical vapour deposition (CCVD) method. Phys Lett 1999;303:117–24.

Nagaraju N, Fonseca A, Konya Z, Nagy JB. Alumina and silica supported metal catalysts for the production of carbon nanotubes. J Mol Catal A Chem 2002;181:57–62.

Nikolaev P. Gas-phase production of single-walled carbon nanotubes from carbon monoxide: a review of the HiPco process. J Nanosci Nanotechnol 2004;4:307–16.

Nikolaev P, Bronikowski MJ, Bradley RK, Rohmund F, Colbert DT, Smith KA, Smalley RE. Gas-phase catalytic growth of single-walled carbon nanotubes from carbon monoxide. Chem Phys Lett 1999;313(1–2):91–7.

Ning Y, Zhang X, Wang Y, Sun Y, Shen L, Yang X, Van Tendeloo G. Bulk production of multi-wall carbon nanotube bundles on sol-gel prepared catalyst. Chem Phys Lett 2002;366:555–60.

Park TJ, Banerjee S, Hemraj-Benny T, Wong SS. Purification strategies and purity visualization techniques for single-walled carbon nanotubes J Mater Chem 2006;16:141–154.

Peigney A, Laurent C, Flahaut E, Bacsa RR, Rousset A. Specific surface area of carbon nanotubes and bundles of carbon nanotubes. Carbon 2001;39:507–14.

Reich S, Thomsen C, Maultzsch J. Carbon nanotubes: basic concepts and physical properties. Wiley-VCH; 2004 ISBN: 3-52740-386-8.

Rohmund F, Falk LKL, Campbell EEB. A simple method for the production of large arrays of carbon nanotubes. Chem Phys Lett 2000;328:369–73.

Saito R. Physical properties of carbon nanotubes. World Scientific Publishing; 1998 ISBN 1-86094-223-7.

Saito R, Dresselhaus G, Dresselhaus MS. Physical properties of carbon nanotubes. Imperial College Press; 1998 ISBN 1-86094-093-5.

Satishkumar BC, Govindaraj A, Sen R, Rao CNR. Single-walled nanotubes by the pyrolysis of acetylene-organometallic mixtures. Chem Phys Lett 1998;193:47–52.

Sen R, Govindaraj A, Rao CNR. Carbon nanotubes by the metallocene route. Chem Phys Lett 1997;267:276–80.

Singh C, Shaffer MSP, Windle AH. Production of controlled architectures of aligned carbon nanotubes by an injection chemical vapour deposition method. Carbon 2003;41:359–68.

Sinha AK, Hwang DW, Hwang LP. A novel approach to bulk synthesis of carbon nanotubes filled with metal by a catalytic chemical vapor deposition method. Chem Phys Lett 2000;332:455–60.

Su M, Zheng B, Liu J. A scalable CVD method for the synthesis of single-walled carbon nanotubes with high catalyst productivity. Chem Phys Lett 2000;322:321–6.

Szabo A, Perri C, Csato A, Giordano G, Vuon D, Nagy JB. Synthesis methods of carbon nanotubes and related materials. Materials 2010;3:3092–140.

Tanaka K, Iijima S. Carbon nanotubes and graphene. 2nd ed. Amsterdam: Elsevier; 2014.

Tang ZK, Zhang L, Wang N, Zhang XX, Wen GH, Li GD, Wang JN, Chan CT, Sheng P. Superconductivity in 4 angstrom single-walled carbon nanotubes. Science 2001;292:2462–5.

Thess A, Lee R, Nikolaev P, Dai H, Petit P, Robert J, Xu C, Hee Lee Y, Gon Kim S, Rinzler AG, Colbert DT, Scuseria G, Tomanek D, Fischer JE, Smalley RE. Crystalline ropes of metallic carbon nanotubes. Science 1996;273:483–7.

Venkataraman A, Amadi EU, Chen Y, Papadopoulos C. Carbon nanotube assembly and integration for applications. Nanoscale Res Lett 2019;14:220. https://doi.org/10.1186/s11671-019-3046-3.

Wang X, Hu Z, Chen X, Chen Y. Preparation of carbon nanotubes and ... chemical vapor deposition. Scr Mater 2001;44:1567–70.

Wang X, Li Q, Xie J, Jin Z, Wang J, Li Y, Jiang K, Fan S. Fabrication of ultralong and electrically uniform single-walled carbon nanotubes on clean substrates. Nano Lett 2009;9:3137–41.

Wei BQ, Vajtai R, Ajayan PM. Reliability and current carrying capacity of carbon nanotubes. Appl Phys Lett 2001;79:1172–4.

Wei BQ, Vajtai R, Jung Y, Ward J, Zhang R, Ramanath G, Ajayan PM. Microfabrication technology: organized assembly of carbon nanotubes. Nature 2002;416:495–6.

Wilson M, Kannangara K, Smith G, Simmons M, Raguse B. Nanotechnology: basic science and emerging technologies. Chapman and Hall; 2002 ISBN 1-58488-339-1.

Yakobson BI, Smalley RE. Fullerene nanotubes: $C_{1,000,000}$ and beyond. Am Sci 1997;84(4):324.

Zaporotskovaa IV, Borozninaa NP, Parkhomenkob YN, Lev V, Kozhitovb LV. Carbon nanotubes: sensor properties. A review. Mod Electron Mater 2016;2:95–105.

Zeng X, Sun X, Cheng G, Yan X, Xu X. Production of multi-wall carbon nanotubes on a large scale. Phys B Condens Matter 2002;323:330–2.

Zhu S, Su CH, Lehoczky SL, Muntele I, Ila D. Carbon nanotube growth on carbon fibers. Diamond Relat Mater 2003;12:1825–8.

Relevant websites

http://nanoscalereslett.springeropen.com.
http://www.azonano.co.
http://www.understandingnano.com.
www.epdf.tips.com.
http://www.nanowerk.com.
http://studentsrepo.um.edu.my.
http://electronicsnode.blogspot.com.
http://fac.ksu.edu.sa.
http://www.intechopen.com.
https://www.cheaptubes.com/carbon-nanotubes-history-and-production-methods-2/.
https://www.cheaptubes.com/carbon-nanotubes-properties-and-applications/.
https://www.sciencedaily.com/terms/carbon_nanotube.htm.
https://www.understandingnano.com/what-are-carbon-nanotubes.html.

Further reading

Al-Rub RKA, Ashour AI, Tyson BM. On the aspect ratio effect of multiwalled carbon nanotube reinforcements on the mechanical properties of cementitious nanocomposites. Construct Build Mater 2012;35:647–55.

Baughman RH, Zakhidov AA, Heer WAD. Carbon nanotubes–the route toward applications. Science 2002;297:787–792.

Carbon Nanotubes—Applications of Carbon Nanotubes (Buckytubes). https://www.azonano.com/article.aspxArticleID=980.

Chen W, Liu Y, Liu P, Xu C, Liu Y, Wang Q. A temperature-induced conductive coating via layer-by-layer assembly of functionalized graphene oxide and carbon nanotubes for a flexible, adjustable response time flame sensor. Chem Eng 2018;J353:115–25.

Durkop T, Getty SA, Cobas E, Fuhrer MS. Extraordinary mobility in semiconducting carbon nanotubes. Nano Lett 2004;4(1):35–9.

Jahanshahi M, Tobi F, Kiani F. Carbon-nanotube based nanobiosensors, In: Paper presented at electrochemical pretreatment. The 10th Iranian chemical engineering conference, Zahedan, Iran; 2005.

Lekawa-Raus A, Patmore J, Kurzepa L, Bulmer J, Koziol K. Electrical properties of carbon nanotube based fibers and their future use in electrical wiring. Adv Funct Mater 2014;24:3661–82.

Purewal M, Hong BH, Ravi A, Chandra B, Hone J, Kim P. Scaling of resistance and electron mean free path of single-walled carbon nanotubes. Phys Rev Lett 2007;98:186808.

Purohit R, Purohit K, Rana S, Rana RS, Patel V. Carbon nanotubes and their growth methods. Procedia Mater Sci 2014;6:716–28. https://doi.org/10.1016/j.mspro.2014.07.088.

Qian D, Wagner GJ, Liu WK, Ruoff RS. Mechanics of carbon nanotubes. Appl Mech Rev 2002;55:495–533.

Schönenberger C, Bachtold A, Strunk C, Salvetat JP, Forr L. Interference and interaction in multi-wall carbon nanotubes. Appl Phys A Mater Sci Process 1999;69:283–95.

Subramaniam C, Yamada T, Kobashi K, Sekiguchi A, Futaba DN, Yumura M, Hata K. One hundred fold increase in current carrying capacity in a carbon nanotube-copper composite. Nat Commun 2013;4:2202–9.

Wan R, Peng J, Zhang X, Len C. Band gaps and radii of metallic zigzag single wall carbon nanotubes. Phys B Condens Matter 2013;417:1–3.

Wilder JW, Dekker C, Venema LC, Rinzler AG, Smalley RE. Electronic structure of atomically resolved carbon nanotubes. Nature 1998;391:59–62.

Zhang Q, Huang JQ, Qian WZ, Zhang YY, Wei F. The road for nanomaterials industry: a review of carbon nanotube production, post-treatment, and bulk applications for composites and energy storage. Small 2013;9:1237–65.

Glossary

Accelerator This increases the reaction rate of *epoxy* systems by adding or producing hydroxyl groups or by increasing the heat produced, thereby increasing the temperature in the system or both.

Additives These are specialist chemicals added to compounds/resins for imparting precise properties, for instance, resistance to UV and catching fire.

Adhesive These are used for mating surfaces for bonding them together by surface attachment. Adhesives are available in liquid, paste, or film form.

Aramid Aromatic polyamide fibers possessing high strength and stiffness.

Blister, blistering Unwanted elevated areas in a molded part created by local internal pressure, because of rapped air, volatile reaction byproducts, or water penetrating by osmosis.

Bulk molding compound (BMC) Also known as dough molding compound (DMC). It contains polyester resin, glass fiber, and diverse mineral fillers. By using BMC in the compression molding process, one can produce high volumes of small complex parts having most favorable corrosion-resistant and electrical properties.

Buckyballs It is the widespread name for a molecule termed Buckminsterfullerene. It consists of 60 carbon atoms formed in the shape of a hollow ball. It was discovered in 1985 by British scientist Harry Kroto. Buckyballs have several fascinating properties. They are very hard to break apart, even at very high temperatures of almost 1000 °C, and would spring back if slammed against a solid thing.

Carbon fiber Carbon fibers are fibers about 5–10 µm in diameter and consist generally of carbon atoms. Carbon fibers possess numerous benefits including high tensile strength, chemical resistance, temperature tolerance, stiffness and low weight, and thermal expansion. More than 95% of all carbon fiber production are applied to composite materials.

Carbon nanotubes Carbon nanotubes have very high elastic modulus and tensile strength. The diameter of most nanotubes is in nanometers. The length of nanotubes can be several million times longer as compared with their width. These are made of hexagonal carbon crystals linked as a hollow tube. Single and multiwall nanotubes have been fabricated.

Catalyst (also called hardener) A catalyst accelerates a chemical reaction, but is not consumed in the reaction. So it remains unchanged and can be recovered at the end of the reaction.

Chopped strands These are short strands cut from continuous filament strands of reinforcing fiber. These are not held together by any means.

Composite It is formed by the combination of two or more materials (selected filler or reinforcing elements and compatible matrix binder) for obtaining particular attributes. Composites are divided into classes on the basis of the structural components.

Compression strength It is the ability of a material to resist loads tending to reduce size, as opposed to tensile strength, which endures loads tending to lengthen.

Core The core material is generally a low strength material. However, its thickness is high, which provides the sandwich composite a high bending stiffness with overall low density. The frequently used core materials are foam, honeycomb, and wood.

Corrosion resistance The ability of a material to resist damage caused by oxidization or other chemical reactions.

Coupling agent These are wide range of chemical compounds. Provide a chemical bond between two different materials, generally an inorganic and an organic.

Cracking Separation of molded material, noticeable on opposite surfaces of a part and extending through the thickness.

Cure Induced by heat or appropriate radiation and may be simulated by high pressure or mixing with a catalyst. It results in chemical reactions causing substantial cross-linking between polymer chains producing an inseparable and insoluble polymer network.

Curing agents Chemical compounds used for curing thermosetting resins. React with primary film-forming matter of coatings and make it solidify to form the film.

Curing time The time taken by a resin to polymerize to its full extent.

Delamination Physical separation between *laminate* plies *through* failure of the adhesive.

Density The density is the ratio of the mass of a substance to its volume. It has units of grams per cubic centimeter.

Direct roving Consist of continuous filaments bonded into a single strand and wound onto a bobbin shape.

Dough molding compound (DMC) It is a dough-like substance consisting of resin, fiber reinforcement, and filler to which pigments and other materials are added.

Electrical conductivity Measures the ability of a material to conduct an electric current. Carbon fiber is an electrical conductor, whereas fiberglass is not.

Fatigue resistance When a material is subjected to cyclic loading, localized and progressive structural damage takes place. Fatigue usually results in disastrous failure, where no clear indication of harm is present when the material fails.

Heat deformation temperature The heat distortion temperature is an index of the strength (elastic modulus) of a material at high temperature.

Fiber Fiber is unit of matter characterized by high length to width ratio, flexibility, and fineness.

Filament A single textile element of smaller diameter and very long length considered as continuous.

Filler Fillers are added to resins or binders that can improve specific properties and make the product low cost or a mixture of both. The largest segments for fillers are elastomers and plastics.

Finishing Application of coupling agent to textile reinforcements for improving the fiber/resin bond.

Flexural strength It represents the maximum amount of stress a material can withstand without breaking.

Flow The movement of a resinous material under pressure, to fill all parts of a closed mold.

Fracture Delamination or cracks resulting from physical damage.

Gel The initial jelly-like solid phase that gets developed during the formation of a resin from a liquid state.

Gelcoat A thin layer of unreinforced resin on the outer surface of a reinforced resin molding. It conceals the fiber pattern of the reinforcement, shields the resin/reinforcement bond, imparts even external finish, and can also impart special characteristics. It is generally pigmented.

Glass fiber The major reinforcement for polymer composites. Reinforcing fiber produced by drawing molten glass through bushings. It possesses superior strength and processability, and the cost is low.

Glass transition temperature (Tg) Tg, not to be confused with melting point (Tm). It is the temperature range where a thermosetting polymer changes from a "glassy" state to a more pliable, amenable, or "rubbery" state.

Graphite fibers A group of carbon fibers having a carbon content of ~99%. This term is used interchangeably with "carbon fibers." These fibers show outstanding intrinsic thermal stability and also have high modulus values. Their mechanical properties **are** retained up to very high temperatures.

Hybrid A composite laminate consisting of laminae of two or more composite. It can also be applied to woven fabrics having more than one type of fiber.

Impact strength It is the capability of the material to endure a load applied suddenly and is expressed in terms of energy.

Impregnate Soaking the voids and interstices of a reinforcement with a resin.

Injection molding Method of forming a plastic to the desired shape by forcing plastic softened with heat into a cooler cavity under pressure or thermosetting polymer **into** a heated mold.

Honeycomb Light weight structure produced from metallic or nonmetallic materials and formed into hexagonal nested cells. The look is alike to the cross section of a beehive. Honeycomb can also be polymer materials in a stiff, open-cell structure.

Laminate Assembly of layers of fibrous composite materials that can be joined to provide desired engineering properties.

Layup A resin-impregnated reinforcement in the mold, before polymerization.
Mat An extensively used sheet-type reinforcement made up of filaments, staple fibers, or strands linked together lightly.
Monomer A compound containing a reactive double bond, able to polymerize.
Polyacrylonitrile (PAN) It is the precursor for carbon fibers used as reinforcement. It is stronger in comparison with pitch-based fibers.
Precursor fibers Raw material for carbon fiber production. These fibers are heated and carbonized in several steps to become carbon fiber. Normally, there are two types of common precursor fibers—PAN and pitch-based fibers.
Polyester Usual term for an unsaturated polyester resin. Unsaturated polyester resins are referred to "polyester resins" or simply as "polyesters."
Polymer A polymer is a long-chain molecule consisting of a large number of repeating units of identical structure.
Porosity Porosity is the quality of being porous or full of tiny holes. Liquids pass right through things that are porous.
Postcure Postcuring is the process of exposing a part or mold to high temperatures to speed up the curing process and to take full advantage of the physical property material. This is typically done after the material has cured at room temperature for at least few hours.
Preform Reinforcement preshaped to the general geometry of the proposed molded part. It is used on more complex and deep-draw moldings for optimizing distribution and orientation of fibers.
Pre-preg "Preimpregnated" composite fibers where a thermoset polymer matrix, for instance, epoxy or a thermoplastic resin, is already present. The fibers usually take the form of a weave, and the matrix is used to join them and to other components during production.
Reactive resins These are liquid synthetic resins that are cured by polymerization or polyaddition to thermosets or elastomers. Curing takes place under low heat and low atmospheric pressure. Typical examples of reactive resins are epoxy resins and polyurethane.
Release agent Prevents other materials from bonding to surfaces. It provides a solution in processes involving mold release, die-cast release, plastic release, adhesive release, and tire and web release.
Reinforcement Process of reinforcing or strengthening. Key element added to resin for providing the desired properties. It varies from short and continuous fibers through complex textile forms.
Resin Polymer with indefinite and often high molecular weight and a softening or melting range that shows a tendency to flow when put through to stress. As composite matrices, resins bind **together** the reinforcement material in **composites**.
Resin, epoxy It has very high tensile strength, stiffness, and fatigue resistance of the commonly used resins. Epoxy resins have performance advantages over polyester and vinyl esters.
Resin, vinylester It is similar in use to polyester. Characteristics of vinylester lie between polyester and epoxy in stiffness, strength, and fatigue resistance and is also in the middle for adhesive properties and price.
Resin transfer molding (RTM) It is a closed mold process for producing high-performance composite components in medium volumes.
Roving A number of strands, tows, or ends collected into a parallel bundle without deliberate twist.
Sandwich structure Composite consists of lightweight core material to which two moderately high strength, thin, dense, functional, or decorative skins are attached.
Sheet molding compound (SMC) A flat pre-preg material, consisting of thickened resin, glass fiber, and fillers. These are covered with polyethylene or nylon film on both the sides, ready for press molding.
Specific strength Strength divided by weight per unit volume. Light and stronger materials have a high strength-to-weight ratio.
Stiffness The stiffness of a material is *measured* by its modulus of elasticity. Carbon fiber has high modulus.
Strand An assembly of parallel filaments produced simultaneously and lightly bonded.
Tensile strength The maximum stress that a material can resist when stretched or pulled before necking.

Thermoplastic Thermoplastic pellets soften when heated and become more fluid as more heat is administered.

Thermoset A polymer that is hardened irreversibly by curing from a soft solid or viscous liquid prepolymer or resin.

Tow or roving An untwisted bundle of continuous filaments. Tows are designated by the number of fibers they have. Tow is used for wrapping shapes or providing remarkable increase in strength in a layup when applied in a specific direction.

Wetout Complete saturation/wetting of a fibrous surface with a liquid resin.

Young's modulus Measures the stiffness of a solid material. It is the ratio of the applied *stress* **to** the *strain*.

Relevant websites

https://www.fibreglast.com/product/glossary-of-terms-in-composites/Learning_Center
https://compositesuk.co.uk/composite-materials/glossary-terms
www.christinedemerchant.com
fibreglass.com
ethw.org
www.compositespress.com
mordorintelligence.com
en.unionpedia.org
pt.scribd.com
composite.com.au

Abbreviations

ACM	advanced composite material
AFRA	Aircraft Fleet Recycling Association
AFRP	aramid fiber-einforced polymer
BMC	bulk molding compound
CFRP	carbon fiber-reinforced polymer
CNTs	carbon nanotubes
DMC	dough molding compound
ELV	end-of-life vehicle
FBP	fluidized bed process
FRP	fiber-reinforced polymer
GFRPs	glass fiber-reinforced polymers
IM	injection molding
NGV	natural gas vehicle
PAMELA	Process for Advanced Management of End-of-Life Aircraft
PAN	polyacrylonitrile
PEO	polyethylene oxide
PET	polyethylene terephthalate
PP	polypropylene
Prepregs	preimpregnated materials
RCF	recycled carbon fiber
RCFRP	recycled carbon fiber-reinforced polymer
SCFS	supercritical fluids
SMC	sheet molding compound
TMC	thick molding compound
VCF	virgin carbon fiber

Index

Note: Page numbers followed by *f* indicate figures and *t* indicate tables.

A

Acetic acid pulping process, 28–29, 54*t*, 60–61
Acetylation process, lignin, 28
Acid treatment, 203
Acrylonitrile, 15–16, 34, 162.
    See also Polyacrylonitrile (PAN)
Adherent Technologies Inc., 100–103, 100*f*, 102*f*, 106–107, 126–127
Advanced Composites Group, 134
Aeronautics industry, 162–164
Aerospace and defense, 2–3, 5, 7–8, 143–144, 161, 171–173, 178
Aftermarket automotive, 140–141
Airbus A350, 91
Airbus SAS, 151
Aircraft, 1, 4, 150–151, 171–173
Aircraft Fleet Recycling Association (AFRA), 106–107, 150–151
Aksa Akrilik Kimya San A, 159–161
Alcell lignin, 29, 34, 60
Aligned TS, 164, 166–167*t*
AMC-8590-12CFH, 142
American Plastics Council, 101–103
Annealing, 203
Arc-discharge method, 193–194, 193*f*
Armchair carbon nanotube (CNT), 190
Aromatic biopolymer, 8–9
Atmospheric Glow Technologies, 175
Atmospheric pressure plasma process, 175
Audi, 174
Automobiles, 140–143, 177
Automobili Lamborghini, 130*t*
Automotive industry, 141–142
    carbon fiber, 178
    recycled carbon fiber-reinforced polymer, 164
Axial compressive strength, 84–86
Axial tensile property, 77, 79*t*

B

Bell Helicopter Textron Canada, 130*t*
Biocompatibility, 148–149
Biolignin TM, 60–61
Biomaterials, 8, 148

Biopolymer, aromatic, 8–9
Biotechnology, carbon nanotube, 199
Bluestar Fibers Company Limited, China, 159–161
BMC. *See* Bulk molding compound (BMC)
BMW, 2–3, 117, 161, 174, 180
Bobbins, 18
Boeing Company, 151
Boeing 787 Dreamliner, 91, 151, 163–164
Boeing's Dreamliner aircraft, 144
Brunauer-Emmett-Teller (BET) surface area, 16–17
Buckminsterfullerene (buckyball/C60), 187–188
Buckypapers, carbon nanotube, 198
Bulk molding compound (BMC)
    CFRP, 95–97
    press molding of, 126
    recycled carbon fiber, 126–127, 136*t*

C

Carbonaceous pitch, 20
Carbon Fiber Flocks, 117
Carbon fiber-reinforced carbon composites, 2*t*, 86, 87*t*
Carbon fiber-reinforced polymer/plastics (CFRPs), 7, 91, 134–135, 179
    aerospace, 143–144
    applications, 139–140, 158
    in automobiles, 140–143, 177
    chemical recycling, 98–116
    composite market, 171
    construction industry, 145–147
    energy and environmental benefits, 118
    energy consumption of recycling process, 116–117, 116*t*
    marine industry, 147
    mechanical recycling, 95–116
    medical applications, 147–149
    microwave thermolysis, 104, 104*f*
    military, 144
    oxidation in fluidized beds, 108–109, 108*f*, 110*t*
    prestressed concrete structures, 146–147
    pyrolysis, 98–108
    recycling methods, 94–95, 95*f*

Carbon fiber-reinforced polymer/plastics (CFRPs) (*Continued*)
  sporting goods, 145
  supercritical fluid method, 110–111, 111*f*, 111*t*
  thermal recycling, 98–116
  treatment, 91
  wind turbine blades, 140
Carbon fibers, 1, 4–5, 139, 160*t*
  application, 2–3, 3*t*, 6–7, 139–152, 140*t*, 157–158
  axial tensile properties, 77, 79*t*
  capacity, 158, 159*t*
  categories based on modulus, 79, 80*t*
  classification, 7, 77, 79, 81, 81*t*
  commercial, 84, 85*t*, 178
  cost reduction, 183
  demand, 2–4, 3*t*, 5*f*
  diameter, 4
  disadvantage, 8
  engineering materials properties, 86–88, 87*t*
  environmental applications, 151
  estimated global consumption, 4, 4*t*
  in flywheels, 173–174
  future research on, 183–186
  global share by application, 146, 146*t*
  high-strength, 1, 5
  in India, 152
  industrial applications, 162–163
  industries future directions, 171–180, 172*t*
  manufacturers, 158, 159*t*
  manufacturing cost breakdown, 157, 158*t*
  manufacturing facility map, 157, 158*f*
  market demand, 171
  market share depend on precursor type, 5, 6*t*
  mechanical properties, 7, 81, 82*t*, 183
  microstructure, 183
  porosity measurement, 84
  precursors, 13, 38–40, 77
  properties of, 1, 2*t*, 77, 78*t*
  recycling of (*see* Recycled carbon fiber (RCF))
  safety concerns, 40–41
  sizing process, 18
  specific gravity, 77, 78*t*
  tensile properties of, 102*t*
  thermal and electrical conductivity, 86–88, 87*t*
  worldwide production capacities, 158, 159*t*
Carbon Fiber Technology Facility, 179–180
Carbonization, 71, 177
  cellulose-based carbon fibers, 22–24
  kraft lignin fibers, 34–35
  lignin, 29, 39, 56–57
  pitch-based carbon fibers, 20
  polyacrylonitrile, 15–16, 17*f*, 26
Carbon nanofibers (CNFs), 34–35, 148, 192
Carbon nanotubes (CNTs), 37–38, 187, 190–192, 191*t*
  acid treatment, 203
  annealing and thermal treatment, 203
  arc-discharge method, 193–194, 193*f*
  bamboo-type, 200
  biotechnology, 199
  buckypapers, 198
  catalysis, 196–197
  chemical vapor deposition, 195, 195*f*
  with closed ends, 188, 188*f*
  diameters, 187–188
  discovery, 192
  electrodes, 198
  electron microscopy, 205
  fabrication techniques, 192–195
  factors affecting growth, 199–202
  field emission displays, 198
  filtration, 199
  laser ablation technique, 194–195, 194*f*
  materials, 196
  mechanical and thermal properties, 190
  microfiltration, 203–204
  morphological characterizations, 205
  multiwalled (*see* Multiwalled carbon nanotube (MWCNT))
  nanomedicine, 199
  optoelectronic applications, 198
  oxidation, 202–203
  photonic applications, 198
  properties, 187–192, 191*t*
  purification, 202–204
  scanning electron microscopy, 203, 204–205*f*, 205
  sensors, 197–198
  shape, 188, 188*f*
  single-walled (*see* Single-walled carbon nanotubes (SWCNT))
  structural characterizations, 205
  structures, 189–190, 189–190*f*
  thermal and electrical conductivity, 86–88, 87*t*
  transistors, 197
  transmission electron microscopy, 205, 206*f*
  ultrasonication, 203
  uses and applications, 196–199
Catalysis, carbon nanotube, 196–197
Cellulose-based carbon fibers (CBCF), 22–24, 25*f*

Cellulosic fibers
  mechanical properties, 81, 82t
  precursors, 81, 83t
CFK Valley Recycling, 117
CFK Valley Stade Recycling GmbH, and Company, KG, 106
Chalmers University of Technology, 149
Chemical degradation treatment, 115
Chemical recycling, CFRP, 98, 110–114
Chemical vapor deposition (CVD), 195, 195f, 201–202
Chirality, 189
Chopped carbon fiber, 163–164
Classical laser ablation technique, 194, 194f
CNFs. See Carbon nanofibers (CNFs)
CNTs. See Carbon nanotubes (CNTs)
Commercial carbon fiber, 84, 85t, 178
Compagnie Industrielle de la Matiére Végétale (CIMV) process, 60–61
Composites., 1, 4–7, 139–140 See also specific composites
The Composites Group, 142
Compression molding, 126–128, 136t
Compressive strength, 84–86, 133–134
Coniferyl alcohol, 52, 52f
Construction industry, 145
Corporate Average Fuel Economy, 26–27
Cost reduction, carbon fibers, 157, 158t, 183
p-Coumaryl alcohol, 52, 52f
Crude lignin, 27–29
Cytec Industries, 159–161

D

Dalian Xingke Carbon Fiber, 159–161
Defense and aerospace, 2–3, 5, 7–8, 143–144, 161, 171–173, 178
Department of Energy, 142
Direct melt spinning, 69–70
Direct molding, 136t
DowAska, 162
Dow Automotive Systems, 174, 180
Dry-jet wet spinning, 68–69
Dry spinning, 67
DuPont, 13

E

Ehime University, 130t
Electrical conductivity, 85t, 86–88, 87t
Electrodes, carbon nanotube, 198
Electron microscopy, carbon nanotube, 205

Electrospinning, 34–35, 69–71
ELG Carbon Fiber Ltd., 106–107, 106t, 107f, 117, 130t, 163
ELG Haniel of Germany, 106
End-of-life (EoL) Vehicle Directive, 92, 133–134
Engineering materials, carbon fiber, 86–88, 87t
ERCOM, 95–96, 96t
Ethanol, 28, 51, 54t, 59–60
Europe, carbon fiber in, 141
Exel Composites, 134

F

Fabrication of carbon nanotube (CNT)
  arc-discharge method, 192–194, 193f
  chemical vapor deposition process, 195, 195f
  laser ablation, 194–195, 194f
Farboud, 141
FEDs. See Field emission displays (FEDs)
Fiber alignment, 128–129, 133, 135
Fiber reclamation processes, 98
Fiber-reinforced polymers (FRPs), 179.
  See also Carbon fiber-reinforced polymer/plastics (CFRPs)
Fiber texture, 77–79
FibreCycle project, 129
Field-effect transistors (FETs), 197
Field emission displays (FEDs), 198
Filtration, carbon nanotube, 199
Fisipe-Fibras Sinteticas de Portugal, 159–161
Flexural strength
  pitch-based carbon fibers, 183
  recycled carbon fiber-reinforced polymer, 133–134
Fluidized bed process, 127–128
  carbon fiber-reinforced polymers, 108–109, 108f, 110t
  carbon nanotubes, 202
Ford company, 174, 180
Ford, Henry, 185
Formic acid, 54t
Formosa Plastics Corp., 173
Free electron laser (FEL), 194
Fuel-efficient automobiles, 178
Fullerenes, 187

G

Gas-phase grown carbon fibers, 36–37
Glass fiber-reinforced polymer (GFRP), 128–129
Glass transition temperature, 69 70
Global Market Insights Inc., 150

Graphene, 187
Graphite fiber. *See* Carbon fibers
Graphitization, 72
Grass lignin, 52. *See also* Lignins
Green Carbon Fibre, 100–101
Guaiacyl (G), 52, 52f

H

Hadeg Recycling Limited, 106
Hardwood lignin, 27, 52. *See also* Lignins
   Alcell *vs*., 29, 34
   characteristics, 52, 54t
   chemical structure, 52, 53f
   problems with, 36
   spinning, 73
Heat treatment, 6–7
Hexcel Corporation, 159–162, 173, 180
High-performance discontinuous fiber (HiPerDiF) method, 132
High-strength carbon fiber, 1, 5
Honda Motor Company, 151–152
Hydrochloric acid, 59, 203
Hydrodynamic alignment method, 133
Hydrogenation process, 27
Hydrogenolysis of lignin, 58–59
Hydrotropic lignin, 61
Hyosung, 162

I

Imperial College of London, United Kingdom, 130t
India, carbon fiber in, 152, 176–177
Indulin AT, 29
Injection molding, 126–127
Institute for Advanced Composites Manufacturing Innovation (IACMI), 142
Interlaminar shear strength, 16–17, 133–134, 183
Invicta, 141
Isotropic pitch, 19–20

K

Kangde, 162
Kayocarbon fiber, 27
Kemrock Industries and Exports Limited, 159–161, 179
Kraft lignin, 56–58. *See also* Lignins
   characteristics, 52, 54t
   melt spinning, 28
   physical appearance, 61, 62f
   problems with, 35–36
   production, 31–33
   sulfur content and purity of, 61, 62t

L

Lamborghini Murciélago, 141
Laser ablation technique, 194–195, 194f
Lignifications, 51–52
Lignins., 8–9, 25, 51, 176, 184 *See also specific lignins*
   acetic acid pulping, 28–29, 54t, 60–61
   advantages, 9, 10t, 25, 26t
   application, 51
   characteristics, 52, 54–55t, 55
   composition, 52
   fractions, 74
   hydrogenolysis, 58–59
   hydrotropic process, 61
   manufacturers, 55, 55t
   mechanical properties, 31, 31t
   phenylpropane units, 52, 55
   physical appearance, 61, 62f
   physical properties, 28, 30–31, 30t
   and polyethylene oxide, 29–30, 30t
   problems with, 35–36
   production, 51
   production steps involved in carbon fiber, 70, 70f
   scanning electron microscopy, 32, 32f
   spinning techniques, 67–71, 68t
   sulfur content and purity of, 61, 62t
   types of, 27–29, 28t
LignoBoost, 31, 57
Lignosulfonates, 55–56
   characteristics, 52, 54t
   physical appearance, 61, 62f
   sulfur content and purity of, 61, 62t
Liquid fuel, 177
Lithium-ion battery, 177–178
Lola group, 130t
Low modulus carbon fiber, 79
Low-reinforced TS, 164, 166–167t
Low-temperature, low-pressure recycling reactor, 101–103, 103f
Luleå University of Technology, 130t

M

Maleic anhydride-modified polypropylene (MAPP), 126
Marine industry, 147

Materials Innovation Technologies (MIT), 93–94, 107–108, 126–127
Materials Innovation Technologies RCF (MIT-RCF), 106
McLaren Mercedes SLR, 141
MeadWestVaco Co., 56–57
Mechanical properties, 7, 81, 82t, 183
　carbon nanotubes, 190
　cellulosic fibers, 81, 82t
　lignins, 31, 31t
Mechanical recycling, CFRP, 94–97
　applications, 97
　thermal/chemical process-based recycling, 98–116
Medical applications, 147–149
Medium reinforced TS, 164, 166–167t
Melt spinning, 67, 176, 184
Mercedes-Benz, 174
Mesophase pitch, 183–184. *See also* Pitch-based carbon fibers
　production, 19–20
　spinning temperature, 20
MG Rover, 141
Microfiltration, 203–204
Microwave thermolysis, 104–105, 104f
Military, 4–5, 144, 171–173
Milled Carbon, 99–100, 106–107
Ministry of Economy, Trade and Industry, 151–152
MIT-LLC, USA, 117
Mitsubishi Chemical Holdings Corporation, 159–161
Mitsubishi Rayon Co. Ltd., 144, 151–152, 159–162, 173, 180
Modulus of elasticity, 28t, 30–31, 30t, 79, 80t
Molten salt treatment, 115
Monolignol monomer species, 52, 52f
Multiwalled carbon nanotube (MWCNT), 188–189, 188f. *See also* Carbon nanotubes (CNTs)
　arc-discharge method, 193–194
　discovery, 192
　growth process, 201
　laser ablation, 194–195
　optoelectronic and photonic applications, 198
　transmission electron microscopy, 205, 206f

N

Nanomedicine, carbon nanotube, 199
National Technical University of Athens, Greece, 130t

Natural gas vehicle, 173–174
Natural lignin, 51–52. *See also* Lignins
Net Composites, 134
Nippon Kayaku Co., 27
Nissan Motor Company, 151–152
Nitrogen-doped carbon nanotube (CNT), 199
North Carolina State University, 130t

O

Oak Ridge National Laboratory (ORNL), 6–7, 10, 174–175, 179–180
Oil exploration, 178
One-pot melt spinning technique, 68
Optoelectronic applications, carbon nanotube, 198
Organoclays, 33–34
Organosolv lignin, 28, 59–62, 176, 184. *See also* Lignins
　characteristics, 52, 54t
　physical appearance, 61, 62f
　sulfur content and purity of, 61, 62t
Original Equipment Manufacturers (OEMs), 161–162
Oxidation
　carbon nanotube, 202–203
　in fluidized beds, 108–109, 108f, 110t
　lignin to carbon fiber, 68–69, 71
　pitch precursor fibers, 20, 21f

P

PAN. *See* Polyacrylonitrile (PAN)
Papermaking method, 127–129
Petroleum based precursors, 175
Phenolated lignin, 27–28
Phenolysis process, 27–28
Phenylpropane, 52, 55
Phoenix Fiberglass, Inc., 95–96, 97t
Phoenix reactor, Adherent Technologies Inc., 101–103, 102f
Photonic applications, carbon nanotube, 198
Pitch-based carbon fibers, 19–22
　compressive strength, 84–86
　flexural and shear strength, 183
　high-modulus, 84
　manufacturing process, 19f
　oxidative stabilization, reaction mechanism, 21f

Pitch-based carbon fibers (Continued)
   vs. polyacrylonitrile, 77, 84–86, 184
   scanning electron micrographs, 22f
   structural parameters, 83–84, 84t
   tensile strength, 183–184
   worldwide production capacities, 158, 159t
Pittsburgh Coke and Chemical Company, 36–37
Plastics Corporation, 159–161
Polyacetylene, 39
Polyacrylonitrile (PAN), 1, 4–5, 13–18, 157
   carbonization, 15–16, 17f, 26
   chemical structure, 14f
   coating materials, 18
   compressive strength, 84–86
   crystallite size in, 84
   decomposition, 17, 17f
   manufacturing process, 13, 15f
   melt spinning of, 184
   and mesophase pitch, 183
   vs. pitch, 77, 84–86, 184
   production costs, 9t
   scanning electron micrographs, 18, 18f
   stabilization process, 15–18, 39
   structural parameters, 83–84, 84t
   wet-spinning process, 159–161
   worldwide production capacities, 158, 159t
Polydispersity, 28, 36
Polyethylene, 184
Polyethylene oxide (PEO), lignin and, 29–30, 30t
Porsche Carrera GT, 141
Precursors, 13, 38–40, 77, 81, 83t, 175
Press molding of bulk molding compounds, 126
Pristine nanotubes (PNTs), 205, 205f
Process for Advanced Management of End-of-life Aircraft (PAMELA), 150–151
Property improvement, 183
Protolignin, 60
Purification of carbon nanotube, 202
   acid treatment, 203
   annealing and thermal treatment, 203
   microfiltration, 203–204
   oxidation, 202–203
   ultrasonication, 203
Pyrolysis, 6–7, 115–116
   carbon fiber-reinforced polymers, 98–108
   recovered carbon fiber, 128

Q

Quantum Composites, 142

R

Raw materials, 7, 13, 38–40
   carbon nanotube (see Carbon nanotubes (CNTs))
   cellulose-based carbon fibers, 22–24, 25f
   gas-phase grown carbon fibers, 36–37
   lignin (see Lignins)
   pitch (see Pitch-based carbon fibers)
   polyacrylonitrile (see Polyacrylonitrile (PAN))
Rayon fibre, 23–24, 24f
Recycled carbon fiber (RCF), 125, 163–164. See also Virgin carbon fiber (VCF)
   advantages, 92, 92t
   bulk molding compound, 126–127, 136t
   challenges, 125
   composites remanufacturing from, benefits and problems, 135, 136t
   compression molding, 126–128, 136t
   demonstrators manufactured with, 164, 165f, 165t
   direct molding, 136t
   energy and environmental benefits, 118
   energy consumption, 116–117, 116t
   hydrodynamic alignment method, 133
   industry progress, 93–94, 94t
   injection molding, 126–127
   markets for, 163–167
   for mechanical testing, 129–131, 130t
   microscopic analyses, 134
   nonstructural applications, 150–151
   press molding of bulk molding compounds, 126
   properties, 113
   real part demonstrator and commercially available semiproducts, 129–131, 131t
   remanufacturing processes, 135, 136t
   scanning electron microscopy, 98, 99f
   wet dispersions, 129
Recycled carbon fiber-reinforced polymer (RCFRP), 150
   in aeronautics industry, 150–151
   mechanical performance, 127
   nonaeronautical applications, 150–151
   structural applications, 164, 166–167t
   structural demonstrators, 143
   woven, 127–129, 133–135, 136t
Recycled Carbon Fibre, Limited, 100–101
Recycling process, 94–95, 95f
   chemical, 98–116
   energy consumption, 116–117, 116t
   mechanical recycling, 95–116

thermal, 98–116
Reinforced TP, 164, 166–167*t*
Reliance, 178
Reliance Composites Solutions (RCS), 179
Repap Enterprises, 60
Resin film infusion process, 127
Rock West Composites, 185
RUE-SPA-Khimvolokno, 24

S

Saxon Textile Research Institute, Germany, 130*t*
Scanning electron microscopy (SEM)
    carbon nanotube, 203, 204–205*f*, 205
    cellulose-based carbon fibers, 24, 25*f*
    lignin, 32, 32*f*
    pitch-based carbon fibers, 22*f*
    polyacrylonitrile, 18, 18*f*
    rayon-based carbon fibers, 24, 24*f*
SCF. *See* Supercritical fluids (SCF)
Sensors, carbon nanotube, 197–198
SGL Carbon SE, 24, 117, 159–162, 173–174, 180
Sheet molding compounds (SMC), 95–97, 126–129
Sigmatex, 134
Sinapyl alcohol, 52, 52*f*
Single-walled carbon nanotubes (SWCNT), 188–189, 188*f*. *See also* Carbon nanotubes (CNTs)
    arc-discharge method, 193–194
    flexible hydrogen sensors using, 198
    growth process, 199–200
    laser ablation, 194–195
    optoelectronic and photonic applications, 198
    thermal and electrical conductivity, 86–88
Small Business Innovation Research (SBIR) project, 117
Soda lignin, 58
    physical appearance, 61, 62*f*
    sulfur content and purity of, 61, 62*t*
Softwood lignin, 52. *See also* Lignins
    acetylation method, 28
    characteristics, 52, 54*t*
    chemical structure, 52, 53*f*
    spinning, 73
Solidification, 68
Solution polymerization, 15–16
Spacecraft, 1, 4
Specialty Materials Incorporated, 159–161
Specific gravity, 77, 78*t*
Spinning methods, 68*t*, 176
    dry-jet wet spinning, 68–69

    dry spinning, 67
    electrospinning, 69–71
    melt spinning, 67
    wet spinning, 68
Sporting goods, 145
Stabilization process
    kraft lignin fibers, 34–35
    pitch precursor fibers, 20, 21*f*
    polyacrylonitrile, 15–18, 17*f*, 39
Steam explosion lignin, 27, 58–59
Supercritical fluids (SCF), 110, 111*f*, 111*t*
Surface treatment process, 16–18, 72
SWCNT. *See* Single-walled carbon nanotubes (SWCNT)
Swerea SICOMP, 130*t*
Syndiotactic 1,2poly(butadiene) (s-PB), 39–40
Synthetic polymer, 30, 38–39, 73

T

Takagi Seiko Corporation, 151–152
Takayasu Recycled Carbon Fiber (TRCF), 107–108, 134–135
Teijin Limited, 151
TenCate Advanced Composites Holding, 162
Tensile energy absorption, 79–80
Tensile properties, 77, 79*t*, 102*t*
Tensile strength, 23, 28, 28*t*, 31*t*, 80–81, 149–150
    longitudinal, 145–146
    pitch-based carbon fibers, 183–184
    properties, 79*t*, 82*t*
    recovered carbon fiber, 112
Textiles Intelligence, 6–7
Thermal chemical vapor deposition (CVD), 201
Thermal conductivity, 84, 85*t*, 86–88, 87*t*
Thermal properties, carbon nanotubes, 190
Thermal recycling, CFRP, 98
    oxidation in fluidized beds, 108–109, 108*f*, 110*t*
    pyrolysis, 98–108
Thermal shock treatment, 115
Thermal stabilization, 29
Thermal treatment, carbon nanotube, 203
Thermochemical methods, 129–131
3-D engineered preform (3-DEP) process, 117, 128–129
Tilsatec, 134
Titan Technologies, 101
Toho Tenax Company Limited, 159–162, 173
Toray Industries Inc., 6–7, 130*t*, 144, 151, 159–163, 173–174
Toyobo Company, 151–152

Transistors, carbon nanotube, 197
Transmission electron microscopy (TEM), carbon nanotube, 205, 206f
Trek Bicycle, 107–108
TVR, 141

U

Ultimate tensile strength, 80–81, 149–150
Ultrafast laser evaporation, 194, 194f
Ultra-high modulus carbon fibres, 79
Ultrasonication, 203
Umeco Composites Structural Materials, 130t
Université du Quebec Montrea, 130t
University of Leeds, 134
University of Nantes, 130t
University of Nottingham, 108, 118, 130t
University of Seattle, 130t
University of Ulster, United Kingdom, 130t
University of Warwick, 130t

V

Vapor phase deposition, pyrolytic carbon, 16–17
VCF. See Virgin carbon fiber (VCF)
Vestas, 140
Virgin carbon fiber (VCF), 97–98, 125, 134. See also Recycled carbon fiber (RCF)
  benefits, 161
  cost of, 118
  recycled carbon fiber vs., 118
Virgin carbon fiber-reinforced polymer (VCFRP), 129
Voith, 174

W

Wet-spinning process, 67–68, 159–161
Wind power, 177
Wind turbine blades, 7–8, 140
Wood-based biorefinery, 175
Woven TS, 164, 166–167t

X

X-ray photoelectron spectroscopy (XPS), 36, 147–148, 205

Y

Yarn-spinning method, 129
Young's modulus of carbon fibres, 39, 109

Z

Zoltek Companies, 159–163, 173

Printed in the United States
By Bookmasters